ARMS

Allocated Risk Management System

Design Improvement Through

GUI Risk Reduction

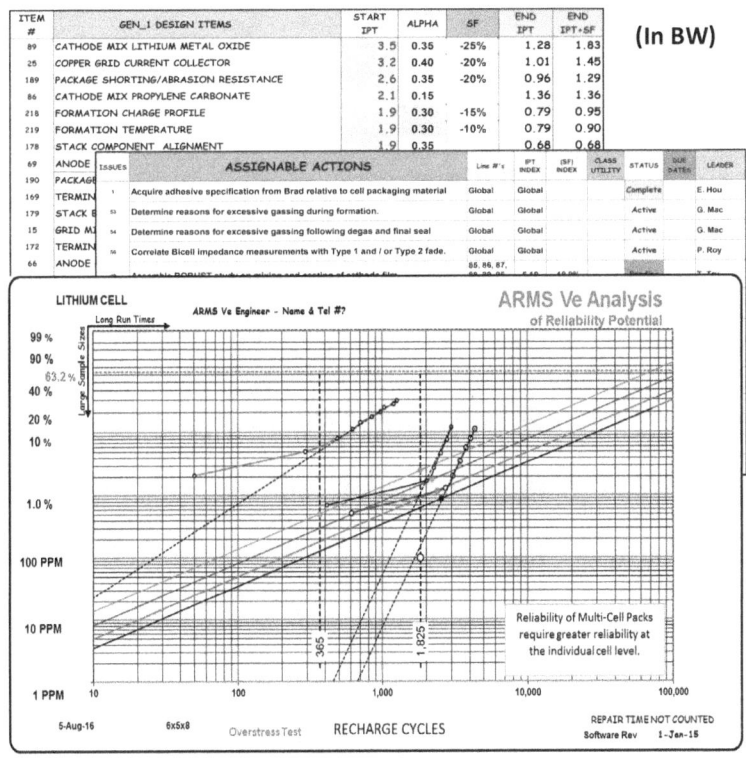

Facilitating Outstanding Leadership & Teamwork

Bill Simpkins

Sidney, Ohio

ARMS training, software and support
is available from:

Bill Simpkins

SPS

8286 Hardin-Wapakoneta Road

Sidney, Ohio 45365

bill9220@gmail.com

(937) 492-9220 Office

ISBN 13:978-1541015494   (2017)

ISBN 10:1541015495   (2017)

ARMS-Publication-171015_BW.pdf

To

Patricia Ann Simpkins

and our Wonderful Children

Patricia "Anne"

William "Bill"

Veronica "Bonnie"

Michael "Mike"

Mary

Carolann "Carolann"

Sharon

John

Susan

Jacqueline "Jackie"

Capt. Guy Gruters,

guygruters@earthlink.net

Motivational, Inspirational Speaker,

Fighter Pilot, "Misty" Pilot, Hanoi POW 5 yrs.

A True American Hero

whose support and encouragement

made this book possible.

Thank you!

To the hundreds of Engineers who supported ARMS by contributing to ARMS' Workshops around the world.

To those who promoted ARMS for their projects.

Dr. Jean Botti,

Ben Gidwani, John Dukro, Larry Oswald, Russ Bosch, Dean Kaman

Bob Burns, Rich Henry, Dr. Gary Cameron, Scott Bailey, Jim Zizelman, Nady Boules, Nick Jones, Walter Piock, Todd Brodewyk, A. J. Lasley, Mark Depoyster, David Quinn, Simon Hudson, Guy Hoffmann, Joseph Bonadies, Steve McMullen, Mike Seino, Mike Frick, Francisco, Ernst Baumgartner, Sanchez, George Simopoulos, Mike Faville, Dr. Sebastian Schilling, Ernst Baumgartner, Martin Hardy, Tarek Kitouni, Joachim Vandulet, Harry Husted, Kaius Polikarpus, and the Engineering Teams under your leadership.

And

Dr. W. Edwards Deming, Dr. Dimitri Kececioglu, Shin Taguchi for your encouragement and technical contributions.

Table of Contents

Preface

ARMS was initiated in late 1991, to meet a need for "outstanding company wide teamwork and innovation."

ARMS can be defined and is applied under several labels: "Allocated Reliability Management System," "Allocated Resource Management System," "Allocated Risk Management System," and "Advanced Risk Management System," depending upon where and how it is being utilized.

In all cases, ARMS is directed company wide teamwork in the reduction of risk through coordinated allocation of resources during innovation and advanced engineering development by all relevant personnel.

This ARMS book was therefore written to provide:
1. Information fostering methodical teamwork and risk reduction with ARMS.
2. ARMS "Train the trainer" resource material with software.
3. A companion for participants during applied ARMS' workshops.

Risk Management
ARMS addresses "Risk" in all of its forms, and ARMS has been applied to risk management on many diverse types of projects. ARMS' workshops may be held for any project. Very importantly, the first and subsequent workshops include key representatives from all company organizations bearing on the project, for example, the CEO, CFO, manufacturing, etc.

This book focuses on *the application of ARMS to Innovation and Advanced Engineering Development of technical products.*

From the very beginning and throughout the project,

with ARMS as a structure, engineers identify where risks reside and their relative magnitudes expressed as a percentage (%) of the total design risk. Even initially, the team may not know exactly what or how something will fail but they have a good idea of where the relative risks of mission failure reside. This allows engineering to start reducing the potential of risks at design conception and on to production with quarterly ARMS' workshops. This is the only program that allows you to significantly address risks beginning at conception, even before the design of hardware has been completed.

ARMS Workshop Model

A series of ARMS' workshops may be held for any project that involves risk reduction, and these workshops necessarily bring together the relevant personnel of organizations in risk management. This ARMS book provides the information needed to bring everyone up to speed with the "how to" specifics of the preparatory tasks required and the mechanisms used including use of the key software. It is effectively the "train the trainer" tool necessary for an optimum ARMS.

Engineering Innovation

What is engineering innovation? Where does innovation come from? What do you need before engineering begins? Do innovation and engineering occur concurrently? How do you lead innovation and engineering into production? Who initiates, fosters, and guides innovation? These are questions that many leaders are paid to answer. Often the first approach to innovation is to throw out the "rule book" and get rid of as many restrictions as possible in the hope that new concepts will fill the void. Yet, the fact is that innovation starts before the engineering, and it continues to accompany engineering development throughout the entire process. **Innovation is thus a vision which engineers translate into a new reality.**

With ARMS as a structure, engineers identify where risks reside and their relative magnitudes expressed as a % of the total de-

sign risk. **The team may not know exactly what or how something will fail, but they have a good idea of where the relative risks of mission failure reside. This allows engineering to start reducing the potential of risks at design conception and on to production with quarterly ARMS' Workshops. This is the only program that allows you to significantly address risks beginning at conception, even before the design of hardware has been completed.**

What is engineering innovation? Where does innovation come from? What do you need before engineering begins? Do innovation and engineering occur concurrently? How do you lead innovation and engineering into production? Who initiates, fosters and guides innovation? These are questions that many leaders are paid to answer. Often the first approach to innovation is to throw out the "rule book" and get rid of as many restrictions as possible in the hope that new concepts will fill the void. Innovation starts before engineering and accompanies engineering development. **Innovation is a vision which engineers translate into a new reality. Think of innovation as finding a new way to fulfill a need.**

Once a vision is formed, it needs to be conveyed to a team consisting of engineers and members of management. ARMS captures the essence of the design vision and states it in a manner that can be understood by all team members. Mathematics can clarify and trigger new visions, however, mathematics itself is not the vision. During an engineering process, new scientific principles may be revealed in the form of breakthroughs, or established scientific principles may be applied in new ways. Innovation is an active process of the human mind, and it cannot be programed into a computer with software.

<u>The Mission Statement of the ARMS' process:</u> **Establish a mission statement vision of the task to be accomplished by a team of engineers and managers, then with ARMS support, translate the vision into relative risks and assignable actions which the team can then execute to achieve all solutions required for a successful outcome.**

This approach supports consideration of performance, delivery, costs, administration support and any other factors that are a threat to a successful completion. Therefore, the project mission statement becomes a common letter of intent with all parties, engineering and management. Competitive pressures and revelations normally introduce a set of evolving conditions during development which require periodic mission reviews and updates as required.

ARMS uses mathematics to support a concept vision and to provide direction to the allocation of problem solving resources. ARMS helps in visualizing the engineering team's mindset through the GUI application of Weibull Analysis Techniques to mental processes and measured physical results.

When science does not support a vision, failure results. As an example, in an attempt to develop a new approach to color television in the 1960s, a cathode ray tube was envisioned based on the de Broglie wave characteristics of electrons. Electrons exhibit both particle and wave characteristics at de Broglie frequencies. An innovative vision was formed to construct an electron beam gun where the beams' velocity could be frequency modulated and the position directed. The velocity modulations were to be at de Broglie wave lengths and the beam directed to strike phosphors on a screen which selectively react to different wave lengths. Phosphors would be selected for the desired colors when excited by the electron beam frequency. The vision

9

had three uniform layers of phosphors, rather than individual dots, thereby making large scale screens affordable. At the time, it was believed to be a vision of interest by the Bell Telephone Laboratories, but unfortunately, science did not support the vision. Just because you can think of something does not mean that it is true or that you can make it work. The vision, however, did win me an engineering position at the BTL in Murray Hill, New Jersey. At BTL, another engineer down the hall, with whom I spoke, Dr. Shockley, put forward a vision that a semiconductor could be changed into a conductor with a small triggering voltage on a controlled basis. We all know how the idea worked out; the transistor and micro-electronics were born.

Someone else's vision I worked on as an engineer was to have a submarine fire a torpedo that would breach the sea's surface, initiate firing a rocket motor, fly downrange and separate a warhead which would then proceed on a ballistic path. The warhead would impact the sea surface, go to a predetermined depth and ignite a nuclear warhead, disabling or killing any submarines in the target area. Warhead depth in the water controlled the size of the lethal area. Quite a vision. However, supersonic impact with the water would impart a 7000 G shock wave through the parts which still had to function. Now there is a vision with a boatload of problems to be solved. The project was named Sub-Roc. This required the design of precision DC motor gearheads for very accurate timing systems that controlled arming the warhead in flight and detonating the warhead at the correct depth after water impact. If the warhead flew off course and travelled long, the timers disarmed the warhead before water impact. The arming strategy required seven timers which had to function perfectly and keep time when energized, even after ten years of storage. The science included developing how to test components to 7000 Gs shock and how to measure the shock produced. Science led to placing two small copper balls in a tube and measuring how much deformation occurred when they slammed together. The energy of copper deformation was supported by mathematical calculations. A test bed was obtained by welding two 16" battleship cannon barrels end to end with breaches at the opposing ends. Two 16" diameter plates with coil springs and anvils in-between, were placed in front of the first breach with the back plate restrained by ball detents. High pressure air was built up between the back plate and the breach until the detents released, allowing the back-plate anvil to smash against the front plate anvil, sending them both down the barrel. At the far end, the second breach bled out air to slow the plates down. +7000 Gs shock occurred when the plates smacked together and a negative 750 Gs drag developed as the plates deaccelerated down the barrel. The reaction forces upon firing would rock the barrels back and forth. The double barrel rolled on rails and was restrained to concrete structures by large coil springs and dampers. The tests were conducted with specimens mounted "in line" and in "transverse" positions. The entire assembly was mounted nearly six stories underground at a Naval facility for safety and noise suppression. This worked out so well the test concept was upscaled to accept the entire warhead. The upscaled test bed was a large diameter heavy steel pipe mounted on two standard flatbed railcars restrained by huge coil springs and dampers. When fired, the railcars, which were monitored by closed circuit TV, rolled back and forth on standard gauge rails. My engineering work was with the smaller 16" guns and the Sub-Roc critical timers. As you can imagine, the initial series of tests resulted in a multitude of failures requiring detailed failure analysis and re-engineering of components.

To be successful, it was necessary to understand the physics of shock. Shock comes in the form of a wave that passes through a test specimen's mass. Motion between adjacent members within the test specimen occurs when the shock characteristics of the matting materials do not match. Analysis and test confirmed that aluminum and long glass fiber Diallyl Phthalate (DAP) plastic under compression demonstrate similar shock characteristics. Other studies included DC motor brush materials, 14 bar commutators, ball bearings, "Vee" wound motor windings and special winding impregnation techniques, the development of recess action internal and planetary gears, and the use of special Delrin and stainless-steel gear materials. **"Recess action gears"** have a unique action where contact between gear teeth only occur on tooth exit. This results in a polishing action during use and provides long life with smooth operation. Eventually the Sub-Roc program was a major success and full deployments were made to all nuclear submarines in our fleet. Later non-nuclear spin-off products, using the DC motors and gearheads, were re-engineered for the Apollo and Viking Lander System missions and are now on the Moon and Mars.

The Viking Mars Mission is an excellent example of philosophy and vision preceding science and engineering. When it was realized that the Moon and Mars might be reachable with existing science, the race with Russia was on. Mars presented several unique challenges that made it more difficult than the moon, not including the manned mission aspects. The trip to Mars required surviving extreme low vacuum in route. The trip to Mars using a Hohmann Transfer Orbit would take about 260 days. A 10^{-14} Torr vacuum was expected in route and several mission critical devices would be exposed to these deep space vacuums. During our development, the lowest achieved vacuum in a large test chamber was 10^{-12} Torr after several days of cryogenic pumping. At such low pressures, strange things happen. All metals in an earth environment contain metal oxides on their surface. Exposure to deep vacuum results in these oxides dissipating and exposing bare metal molecules at the surface. If two separate metal parts touch each other under these conditions, they weld together. Several items on the Viking Lander were in jeopardy due to this phenomenon. The boom arm was two flat strips of metal with holes along the edges like a strip of movie film. Each strip had an opposing hump formed down the length of the center between the holes. One humped up and the other down. The strips were forced flat as they were rolled up like a child's Kazoo. When deployed, they fed out between guides to form an arm and rolled back when not needed. The steel metal guides were the contact problem. Carbon blocks could not be utilized. Carbon requires water for lubrication, oils would dissipate and possibly foul optics. Metals would weld when the oxides were lost. Ceramics had excessive friction. I proposed a new vision. The applied answer was to use a common earth-bound solution with which I was familiar from working on diesel locomotive lubrication problems, using leaded bronze guides with babit metal impregnation. Let weak welds form between the soft and hard metals, then have enough power in the arm deployment motor to break the welds and use the babit as a dry lubricant. This concept is utilized in locomotive bronze sleeve bearings as a backup to forced oil lubrication. With minor adjustment and 10^{-12} Torr testing the idea was confirmed and the concept is on Mars today. This is an example of the value of having engineers on the project team with diverse backgrounds and experiences.

Steel ball bearings have a similar metal weld problem as the guides after metal oxide loss. In the case of ball bearings, the solution was to evacuate ball bearings in a 10^{-12} Torr vacuum chamber, to evolve off the surface oxides, then, while still at vacuum, introduce mineral oil, which has a long chain molecule. The ends of the oil molecules will bond to the exposed metal molecules. This results in an oil film where an end of the oil string is attached to a metal molecule at the surface forming a mineral oil "mohair" which survives prolonged vacuum exposure.

The point of these examples is that mathematics and science alone would never solve the problems. What is needed is an engineer with base knowledge of chemistry, physics and mechanics forming a vision, then utilizing science and mathematics to solve the problems. Without visions of new solutions, solutions will not occur. This is where philosophy precedes engineering. In these examples and more to come, the internet is a fantastic resource to find related examples that may solve your vision. Programs such as Goldfire Innovator and TRIZ can be of great assistance in locating potential solutions.

Utilizing engineering vision is at the heart of what ARMS is about. ARMS pulls the team together, for what is really a brief period, to review the risks in the project and to form a vision of solution paths. Lead Engineers take the vision and assign project team members to resolve potential solutions. With ARMS, instead of having an individual responsible for forming the vision, the individual has the support of the full project team and the team's larger experience and knowledge base.

To not take the time to do an ARMS' Workshop or an equivalent, if there is one, is the same as not taking time to sharpen your axe.

Many component development projects are used to explain ARMS in this handbook, however, do not read into that fact that ARMS is a component improvement process. It is much more than that. ARMS makes its greatest contribution to large system developments which include sub-systems and critical high technology components. ARMS has supported projects with suppliers spread around the globe. The improved communications via ARMS becomes a strong component of overall improvement.

INTRODUCTION

Confirmation of improved risks normally requires the measurement of durability under stress. However, in some industries, sales actually precede design; e.g., airplanes, rockets, and many other capital intense items developed under contract. Actions which lower risks are not easy to measure and frequently designs are not adequately confirmed, due to small or no sample sizes and long or no test periods. Full demonstration of large systems' reliability is rare. The B58 Hustler fighter bomber was a demonstrated reliability program. It was a valiant effort that ended in an overweight, over cost aircraft with a compromise on performance leading to limited deployment. The Lockheed L1011 commercial passenger jet was highly influenced by demonstrating reliability and requiring tighter component performance requirements when compared to the Douglas DC10. During DC10 development, engineers were encouraged to utilize their judgement. The L1011, which was built to spec., was overweight and expensive. The DC10 was a business success for many reasons, most of which can be traced back to the greater local engineering authority in making decisions. Demonstrating risk improvement requires proving a negative. The lack of failures during tests and tests with limited or incorrect stress can yield false positive results. Build, test and fix cycles is the heuristic engineering approach, that is, trial and error. This is a costly, time consuming method prone to failure in today's fast-moving environment. Something better is needed.

ARMS is the answer to achieving company wide teamwork.

Risk improvement is a "growing" activity, which in theory can be demonstrated through validation studies, simulation and tests. In reality, most programs are not adequately funded with money or time to do the work properly. The true dynamics of risk improvement in general are poorly understood by those making budgetary and delivery commitments. This is, in part, due to a lack of broad based teamwork and applied risk improvement experience. Graduate and post-graduate reliability courses rely heavily on mathematics, which are useful when trying to interpret results, but are not as preemptive as needed, or preemptive at all when dealing with advanced design innovation. They are not management courses.

ARMS is preemptive and requires management participation.

The full engineering team needs to be in a position to contribute to design and business decisions. To be preemptive as managers, we need to know and work with what design engineers are thinking, then plan an approach to the future design process. Respecting the engineering mind is an area requiring attention. Reliability improvement activities require support for risk identification and actions leading to risk improvement. ARMS will ask the engineering team to stipulate where the elements of risk in the design reside, where management support is needed and what should be expected if the support is or is not provided.

ARMS drives issues into the light of day and asks what is needed to obtain solutions in detail. Gaps in design and issues that are being ignored are uncovered for action before they become last minute revelations and budget overruns.

Companies often have design guidelines of dos and don'ts which place limits on innovation and risk improvement. For example, if you are making components for use in a salt water marine environment, you might only use materials that are non-corrosive or proven to be adequately protected. For example, new designs are often required to contain a given percentage of older field proven design elements to limit exposure to new risks.

But what about innovation, new product applications, new industries or new engineering teams working beyond their experience envelope? ARMS will benefit all design work, but ARMS is focused on shedding light on the gray or dark areas of a new design vision.

ARMS is directed at improving the performance of the Engineering Team, Management, and Suppliers. Teams which conduct several ARMS' Workshops become better at their everyday functions. ARMS improves the team's understanding of reliability, risk improvement strategies, and how to provide an input to improving project cost. ARMS' Workshops must be approached with an open mind, free of preconceived notions. Concepts and methods presented here are from engineering, not science, and do not raise expectations to unrealistic levels.

ARMS was derived from national and international experience as an engineer, manager and director in a wide range of industries and product classes that include telecommunications, aviation components, classified NSA hardware, space missions, space optic systems, HVAC compressors, AC and DC electric motors, hybrid electric rail vehicles, electric and hybrid automobiles, a wide range of internal combustion engine components and automotive systems.

Experience with womb to tomb tracking of serial numbers with uncommonly excellent application data feedback for millions of HVAC compressors and systems over decades of field service provided uncommonly detailed failure cause and improvement knowledge. The application of 3D warranty graphics presented in this publication to HVAC provided unparalleled insight into field service results.

The prime objective of ARMS is to provide for an organized approach to risk identification and improvement, engineering development and to provide leaders with the foresight to use risk / reliability improvement as a competitive marketing weapon. Customers seek out and reward companies who can demonstrate that they are thinking and acting smarter than their competition.

When dealing with engineered products, customers look for long

term relationships and buy based on an engineering team's knowledge as well as the products they produce.

Reliability is measured by comparing test failures with application requirements and conditions. Without achieving test failures, reliability is only a guess. By failure, I mean not meeting performance specifications during the usable life period in an agreed application environment. For example, a product may be sold with a one-year life expectation and actually meet all requirements for a year and a half. The product is judged to be reliable by the consumer. The same product sold into the same application with a life expectation of five years would be judged unreliable. If the product meets all specifications for three years, then degrades to 85% performance for an additional two years, it may be sold on the basis of degraded performance and can still be considered a reliable product for five years. This is a common approach with rechargeable batteries. Other approaches, that I do not agree with, can be represented by automobile tires that have two layers of rubber, one hard for long life covered by one soft for improved breaking during qualification tests. Initially passing the specification is not enough, you need a design which will meet all specifications for the life of the product. One manufacturer who used the two-layer design, later corrected the design when the deception became known. They then proceeded to advertise the fact that the tires were safe throughout their life. Of course, the tires' lives were now shorter.

To claim that a product is reliable without defining the terms for judging reliability is meaningless. To design a product without setting documented reliability objectives is meaningless. Agreeing to an objective and not taking positive actions to achieve the objective is especially meaningless. Getting it "Right the first time" applies to more than performance. "Right the first time" requires risk reduction and designed in reliability, not just testing and fixing. Unfortunately, Deming's "burning toast and scraping it" describes most reliability improvement approaches taken.

In today's competitive world you must "design reliability in,"

starting at the concept innovation stage and continuing throughout the design development process.

Customers communicate with each other by the internet, text messages, Google searches, Linkedin and Facebook, to name just a few methods. And, of course, people move between companies. If poor reliability exists, everyone learns about it. Without a strong, clear reliability objective and strategy, it is wishful thinking to expect the right decisions to be made on design materials and configurations necessary to produce a reliable product or service.

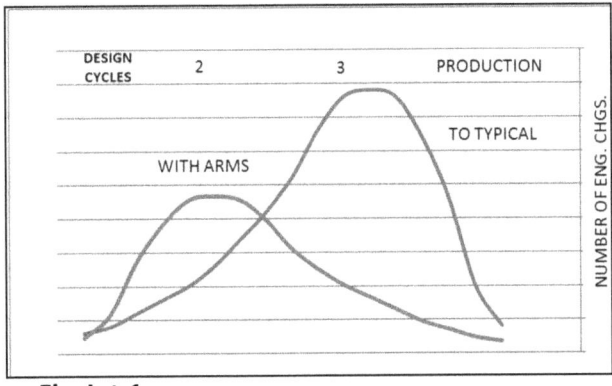

Fig. Int-1

"Product performance" and "cost" are usually easier to measure and become the yardstick to assess engineering performance. Personnel reliability measurements do not exist, even in very sophisticated companies. I cannot think of an example where an engineer's pay was adjusted based on product reliability. Perhaps fired for poor reliability, but how about a raise for good reliability? In some industries, which monitor reliability after products are put into service, reliability requirements are built into product penalty clauses that may impact supplier finances. Major capital industries like rail transport and other governmental capital contracts are examples where an after sale "performance penalty" can have a negative impact. With consumer products, examples of poor reliability are found in the form of warranty payments and/or loss of repeat business.

Known poor reliability may also be covered by better warranties

Contrary to common belief, reliability problems must be addressed during the more innovative engineering development cycles.

Corrective actions close to the start of production are, risky and cause major SOP delays.

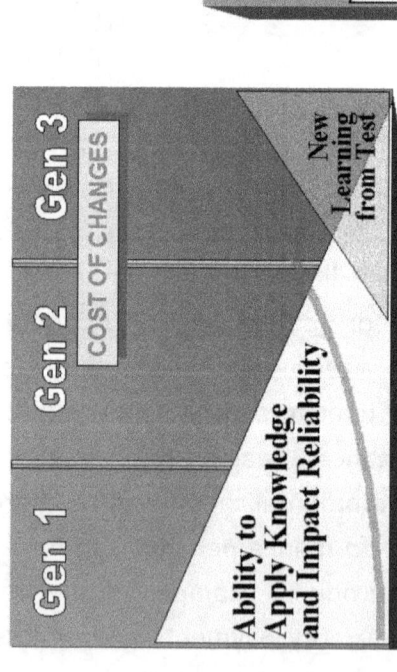

Gen 1 · Gen 2 · Gen 3

COST OF CHANGES

Ability to Apply Knowledge and Impact Reliability

New Learning from Test

Gen 1 · Gen 2 · Gen 3

Knowledge & Experience

Ability of New Learning to Impact Design

Learning from Test

ARMS is focused on solving Reliability issues using Knowledge and Experience at the start of each design phase.

Fig. Int. - 2

to soften consumer pain. I have attended many marketing meetings where warranty costs were estimated and built into the cost of sales. If you see a product with a super warranty and you buy it, you might really need the warranty. Buyer beware. I know people who made purchases based only on best warranty and defend their actions by saying you always need them. If you are really confident that your product has a low risk content, you can offer a better warranty, but that is not the way the market usually works. Monetary reserves to cover estimated risks need to be financially secured. Information from ARMS can help with determining the magnitude of the true financial reserve required.

Reliability improvement is usually in an engineer's internal guidance, but may also be stated in company documents or as a procurement specification. Whether a design actually meets reliability requirements becomes an engineering estimate, determined by limited measurements or through limited application experience gained during normal service usage. There needs to be an exit gate requirement to determine when a product can leave engineering and go into production. ARMS can assist in setting exit gate parameters.

Engineering "knowledge" is gained by engineers learning from observing applied results over time, if they have feedback. Often engineers are not on the job long enough in today's high turnover environment, and leave before observing products in intended applications. The potential knowledge gain is lost by the engineer. The company's investment in the engineer is also lost and full circle design knowledge does not materialize.

Component suppliers may have the ability to monitor a common product in a number of different customer applications. In such cases, reliability data can be of great assistance in helping a customer out of an application problem. In the HVAC compressor industry, supplier supportive actions have great value to the customer.

Design processes to a large extent are "Heuristic," trial and error. A management challenge becomes how to improve design cycles. How can the number of design attempts or cycles be reduced to shorten the time required to capture and accumulate engineering knowledge? Or how can design cycles be faster to get in more design iterations within budget? In a mixed team of engineers (junior and senior), how can a team's engineering knowledge be optimized and shared? Engineers are not known for being open, sharing or receptive to another person's experience. An engineer knows he is considered "smart" until he starts talking. A synergistic framework that allows engineers with diverse levels of experience to communicate without fear of exposure to judgment is needed and is provided by ARMS.

The discussion of reliability improvement requires comprehension of project timing issues. One of the often-claimed characteristics of design is that major efforts to improve a design occur just prior to start of production. Proof of this claim is found in the frequency of engineering changes. I have found this claim to have an element of truth in all industries. Often undocumented changes that occur early in (Design Cycle 1), before formal drawings are generated, are not counted. While many companies do track changes during (Design Cycle 1), they typically have them in "unofficial" and inaccessible files. Even so the curves as shown **(Fig. Int-1)** still apply. Late cycle changes made as a product is being prepared for production carry undesired risks. **Associated late change costs are higher and the probability of having an unproven design produced increase greatly.** Early reliability improvements require making changes from experience when hard information is unavailable, however, that is when change costs are lowest **(Fig. Int-2)**. To the greatest extent possible, reliability improvements need to precede "sunk cost" investments, such as dies, molds and procurement of single purpose machines. Adoption of reliability improvements after major capital intense investments becomes very hard to justify.

The term "Design Cycle" is used loosely in this discussion; you should expect several small iterations of the design or sections of the design within an "Official Design Cycle." As applied here, design cycles usually coincide with budgeted cycles. In the automotive and related industries, on new concepts, it has been shown that, on average, 12 full budgeted design cycles occur before the start of production. Complex concepts, like Solid Oxide Fuel Cells (SOFC), take far more design cycles. Changes adding to a family of existing designs take less, e.g., a new electric motor capability achieved through a programmable winding change.

How can the need for risk improvements be revealed and implemented during the earliest conceptual phases of a design, while innovation is still occurring? If reliability is a measured value, then a physical product is required to be stress tested and break, in order to determine how long we should expect one to last under usage conditions. Normally, stress testing occurs in the third cycle or later when physical test products become available. **Early prototypes are expensive and are not considered representative of the final product in terms of reliability.**

ARMS is an answer to this dilemma and early reliability estimates can be achieved.

The transition to earlier implementation of reliability improvements requires management support for education and the adoption of pre-emptive project engineering processes like ARMS. It is unfortunate, but costs to avoid failures are more difficult to justify than costs to correct failures. **Pre-emptive costs are requested in response to engineering concerns, failure correction costs are in response to a proven reality.** However, if the pre-emptive costs are not applied, higher reality costs are assured. "Do it right the first time" requires pre-emptive efforts during early design with strong management support. Ironically, I have had business managers say they have nothing to do with risk improvement when, in fact, they are the major players. Accounting systems are very poor at assigning the cost of poor reliability back to the engineering team. In fact, the engineering teams that get unreliable designs accepted quickly are rewarded for finishing early and under budget. An inability to assess costs is compounded if responsibility for design is handed off to an independent production engineering group as the design enters preparation for production. Samsung's Galaxy Note 7 lithium batteries seem to be a prime example. The cost of recalling millions of smart phones will be legendary.

ARMS is excellent at identifying the need for improvement and gaining support for early implementation. ARMS will help manage the shift to gain knowledge earlier, as shown in the preceding "Cost of Change" graph, more than any other approach of which I am aware **(Fig. Int-2)**.

When the advanced design work is successful, human nature attributes lower production development costs to the production engineering group. If production design costs are higher than expected and redesign occurs, the excess costs are attributed to the original engineering designers. That's the way the game is played.

ARMS is a management process, involving broad teamwork, that reveals, quantifies and solves design reliability weakness during early conceptual innovation stages and on into production.

The initial "Advanced Design" team requires members with production knowledge.

Early involvement during innovation needs to be an activity for key experienced production personnel.

With management support, ARMS has an equal standing with other business systems during meetings on budget, product cost, engineering performance and schedule attainment. When you have a product with poor reliability, you have a recall economic booby

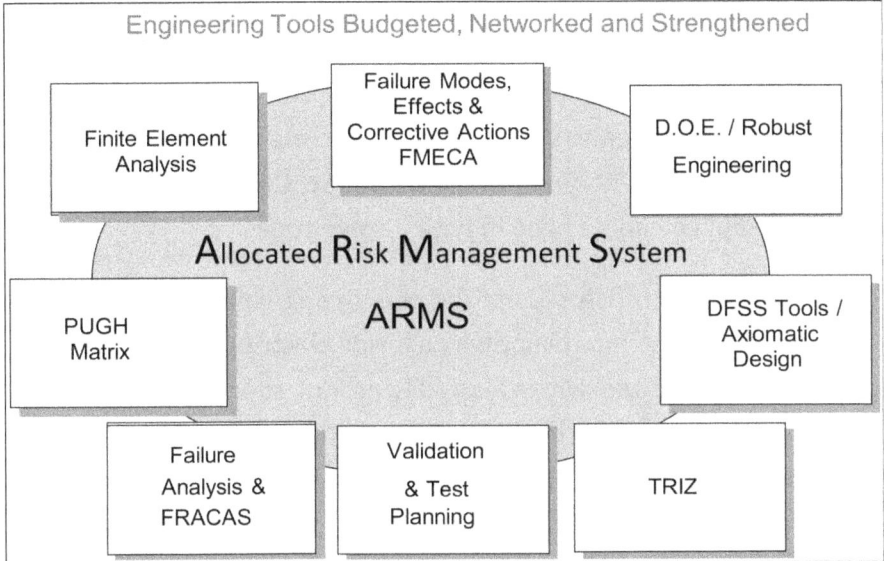

Engineering Tools Budgeted, Networked and Strengthened

Failure Modes, Effects & Corrective Actions FMECA

Finite Element Analysis

D.O.E. / Robust Engineering

Allocated Risk Management System

ARMS

PUGH Matrix

DFSS Tools / Axiomatic Design

Failure Analysis & FRACAS

Validation & Test Planning

TRIZ

Fig. Int.-3

trap. As stated earlier, ARMS was developed by an engineer while working in high-tech industries. In Aerospace, full-up stress testing and postmortem analysis is very expensive and often not possible. "One off" designs, where test hardware is prohibitively expensive and there are application conditions not achievable on earth, are common. ARMS was honed to its present shape through a wide range of applications and several hundred workshops with multinational engineering teams. Applications of ARMS include complex computer controlled electro-mechanical medical implants, people movers, hybrid automobiles, electro-mechanical designs, electronic designs, a wide range of sensors, lithium batteries, fuel injectors, advanced brake systems, steering, hybrid vehicle power electronics, SOFC fuel cells, non-thermal plasma reactors, high tech powered wheel chairs and many other projects. ARMS' models have proven useful in successfully managing engineering budgets while achieving risk reduction objectives. Examples drawn from actual ARMS' Workshops were modified to remove proprietary content and are discussed in later chapters. ARMS has strong positive customer appeal.

As indicated, ARMS allows scheduling and budgeting many other design tools that may be beneficial. The indicated tools are illustrated in ARMS for reference (Fig. Int-3). There are many others that could be added to the chart. A color code system for the added tools is set on the model INPUT sheet. Cells where the tools will be applied can be filled in with a color code.

A Federal DOT office recognized the strong contributions ARMS made during a multi-supplier hybrid electric vehicle project. ARMS was credited with clearly laying out supplier needs, local developments and for leading to elimination of 11th hour surprises that would have surfaced too late for the program to recover within time and budget. **When build and validation test samples are not possible, ARMS makes a particularly strong contribution.** Quarterly full project reviews and focused supplier job sight reviews promote team relationships and internal project transparency. When there is a single customer, inviting the customer to attend an ARMS' Workshop has always proven to be a positive event. Prime customers have utilized ARMS to review their higher level projects.

The capture of knowledge is a strong component of the ARMS' Workshop. Ideally, ARMS is started early in the concept stage of a project. Traditionally, learning from tests occurs late or at the end of a design build and test cycle. With a later start, projects will still benefit from ARMS, but some innovation opportunity is lost. Starting reliability improvement in the early stages makes changes easier to implement at lower costs compared to changes forced by discovery through durability test faults (Fig. Int-2).

During validation durability tests, there are sunk costs in raw materials, tooling, and sub-contracts which inhibit making improvements that require new prototypes or suppliers. Finding problems during testing creates a need for repeating design cycles and tests and also puts extreme pressure on reducing test sample sizes, resulting in lower statistical confidence. Time and money become limiting factors, strongly inhibiting changes. Unless absolutely required, design risks go unresolved. The evaluation of team performance can also suffer from late delivery or no delivery. Unfortunately, late is when the newest knowledge of a need for change comes to the surface. The problem is exacerbated if it is revealed that existing knowledge within the team had not been fully exploited. During the Viking Lander Project, I was the sole reliability consultant assigned to review all supplier hardware designs and validation test results for on-board devices. One supplier of small DC motors was so far behind, failing validation and lacked the expertise required, I had to disqualify the company. I wrapped my supplier review around a bottle of "Viking Cherry Kaffa" wine, which had a label sporting a Viking standing with a long spear dangling a pair of cherries, then gave it to the Viking Program manager. He found a new supplier and qualified new motors just in time.

The ARMS' process is based on a philosophy and supported by mathematics, which aides in shaping an engineering vision.

A strong design philosophy gets us on, and keeps us on, the track to success. Achievement comes from the ability to form a vision of the future, to draw on prior knowledge and to apply mathematical tools following or establishing physical laws. To be successful at engineering or science, we must first educate ourselves, gain a base knowledge of how things work, how materials behave, and then develop a vision of how new combinations of our knowledge will satisfy the new objectives. You have to have the ability to become emotional over a piece of steel (or silicone). This involves hardware and software as well as our understanding of human behavior. The Viking Mars' Lander examples are clear cases requiring the transfer of design visions to new environments. **During innovation and design, an engineer's ability to visualize and undertake actions to bring the vision to a reality are the important tools to be improved.** Mathematical tools are simply aides to quantify and explain the vision, thus they are subordinate to the engineer's mind. Mathematical tools are only of use if there is a vision of where and how to apply them. ARMS recognizes this fact and works to supplement the engineering team's abilities. To a great extent, the mathematics necessary to support ARMS' improvement efforts are built directly into ARMS utilizing Visual Basic, thereby freeing the engineers to focus on creativity. **ARMS provides a GUI interface between the engineering team and the required core statistical equations.** ARMS core mathematics may be found in the ARMS software file "load-arms-ve-2017.xlsm" under an Excel tab with my name on it, "Bill Simpkins."

Using ARMS, a lead engineer can guide the team by drawing on the team's full experience and knowledge. ARMS also produces "cross pollination" between members. Sometimes I hear, "We knew that." Yes, you did, but were you making use of that fact? Were you collecting and evaluating the knowledge within the team or was it a process like osmosis? My wife and I use osmosis, that is, anything she knows I am expected to automatically know as well, and vice versa. If you are married, you know how that works. As engineers, we can do better.

But why ARMS? Why not FMECA or some other tool? Perhaps you have a customer who requires the use of a FMECA or Taguchi Design of Experiments. To start with, ARMS is an Engineering Design Management System, not a specific problem-solving tool. ARMS is about the team and how they perform. ARMS is directed at supporting the minds of the engineering team regarding how to improve engineering performance. ARMS is effective at managing resource allocations, determining when and where other specialized engineering tools will be of benefit, how to plan for budget conflicts and communications across levels of management and cells of engineers. ARMS brings focus to the engineering effort and sheds light on where risks live.

ARMS is an action that you undertake to be more effective than your competition. It is an action that will result in faster design cycles with more iterations per dollar. ARMS develops a vision that you can share with customers and your supply base.

With ARMS, suppliers and customers can participate in the development of complex systems. This is a very powerful Engineering and Marketing tool.

Additionally, my experience in a wide range of industries indicates a systemic problem with FMECA of late utilization. With a FMECA, if a serious problem is revealed, last minute changes result, followed by expedited retesting and a sharp increase in risk and cost for the company, suppliers and the customer. FMECA have great value, but they are a weak safety net. In my experience, FMECA have little positive effect on team dynamics and future performance. They can be useful, but are never timely or enough. FMECA meetings are events, rather than a management process.

As previously stated, ARMS is the outgrowth of many years work across a wide spectrum of industries and products costing from a few dollars each at volumes of over a million a week, to one-off products like orbital KH11 cameras and Viking Landers costing tens of millions of dollars. In the high-volume world like HVAC compressors, if you do not have a reliable product, the failure volume will eat you alive. With the high value aerospace products, without reliability, you have a great, but useless, fireworks display or an expensive orbiting garbage can.

Chapter One

Concepts Utilized within

ARMS

- **Facilitating Outstanding Leadership / Teamwork between all departments**

- **Engineering Design & Innovation Under Control**

- **Investments into improving process capabilities will transfer to the next project. Investments into large sample sizes to be destroyed become waste.**

Before proceeding deeper into ARMS, we need to discuss reliability improvement tools and bring out some concepts that are crucial to understanding and applying ARMS.

Dr. Enrico Fermi Solutions

Dr. Enrico Fermi is credited with the first experiment to have a controlled nuclear reaction. This was conducted, in secret, under Old Stagg Field Stadium in Chicago Dec. 2, 1942.

Dr. Fermi was also famous for a method of solving problems when very little was known about the subject. ARMS will employ Fermi solution techniques. The following illustrates how his process works.

Consider how many piano turners would be needed to provide service to a given Metro area? To estimate a solution, we first set up a model.

How many homes in the Metro area?
How many homes have a piano?
Quantity of pianos in the area?
Years between tuning?
How many to be tuned this year?
Working days per year?
How many need tuning each working day?
How many are tuned per person per day?
Tuners required +/- about X.

A small group of people who live in the Metro area can be asked to estimate answers for each of the questions and arrive at a reasonable answer with very limited access to hard data. This is a Fermi class model.

Fig. 1-1

ARMS risk allocations are part of a Fermi class solution.

With a Fermi Model, you make a first pass at input data estimates. Each line of the model is discussed by the team to obtain their best-informed guess. Then work is commenced to improve the estimates.

For example, for the first question, "How many homes are in the

Metro area?" One or more team members may have a good idea. They may have read about metro homes in a newspaper or seen it on T.V. Perhaps they found the answer using a smart phone or P.C. After a reasonable value is agreed upon, the data is entered. Each line of the model is treated in a similar manner until an initial estimate is obtained **(Fig. 1-2).** After we have an estimate, individuals can be assigned to obtain better answers, for instance a check with City Hall for a census number or the Tax Department for their count. A piano company might be contacted for their estimate of the number of pianos. A member might check with a piano turner for their estimate of time between tunings. Some pianos are played by gray haired ladies at church, while others at the corner bars are pounded on every night playing Scott Joplin, but perhaps nobody notices when they go out of tune. Some wood harp framed pianos can be affected by humidity. What are the weather conditions in this area? Perhaps a sampling survey could be conducted? There are many activities that can be considered during this process. A record needs to be made of the considerations that influenced the team.

We will similarly be making many estimates during the ARMS' program. The more estimates there are, the greater the degree to which we will be tapping into the combined engineering knowledge base and the more accurate our estimate will become. So it is the same process with ARMS, first agree on model input, then make rough estimates, then improve the estimates as design progress occurs and knowledge expands. Capture the reasoning for setting model values in ARMS with the Excel com-

Dr. Enrico Fermi Solutions	
How many Piano Tuners are in the Metro area?	
400,000	Homes
1/20	Have a Piano
20,000	Pianos
5	Years between tunungs
4,000	This year
200	Working days/year
IF 20	Tuned / day / person
10	Tuners +/- about 3 required

With many elements in the model, errors tend to ave out.
Dr. Fermi-Physicist credited with the 1st nuclear chain reaction in a lab.

Fig. 1-2

ment function, document issues that arise requiring solution and then build an action plan to lower risks. Our piano question was initially how many tuners are needed. The model can be improved by including how many turners work full time or part time as a sideline. Building an ARMS' model takes a few days, depending on complexity and innovation content, and "like any engineering model," requires concentrated focus from qualified engineers providing a sincere input. All engineering models by definition are subject to garbage in, garbage out, so sincerity of input is paramount.

At one point in my career, I was asked to be the move coordinator for moving a company from Racine, Wisconsin, to Ft. Lauderdale, Florida. A primary task was to estimate the weight of all the machines and tools in a large machine shop that covered three floors of a large building. By estimating some of the smaller objects, verifying a few with actual measurement and by conferring on many other estimates, we were able to "calibrate" the total estimate. In the end, much to the amazement of everyone, myself included, when the dozen 18-wheel trucks were weighed, the total load was within 1% of the estimate! I am sure that would be tough to repeat. However, this example proves the potential of our Dr. Fermi model thinking in an ARMS' Workshop.

Dr. Hermann Ebbinghaus Curve of Recall (1885).

Repetitive exposure to a concept strengthens our ability to retain information and obtain knowledge. Limited exposure, such as one reading or one education session, leads to rapid memory loss. But frequent exposure can build on a base of knowledge. I will refer to the Ebbinghaus effect often. An example would be to view a movie three or more times over a couple of months. With each viewing, you see things that you missed in the earlier

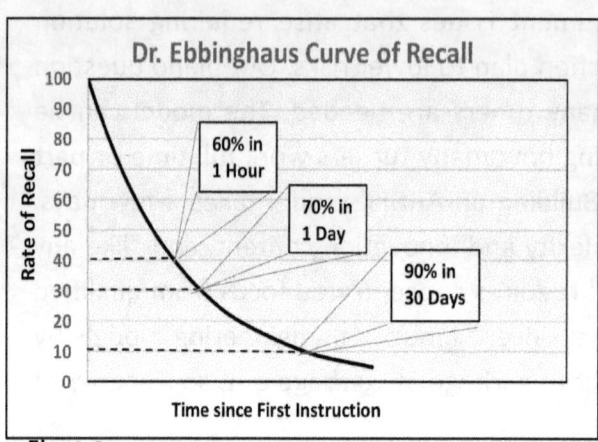

Fig. 1-3

viewings. Shifts in attention to detail explain some of it, but also it is because you forgot some of the lesser things between viewings.

The work of Dr. Ebbinghaus has been studied for many years at the university level around the world. An internet search will provide several studies. His theories have held up in case after case. The Curve of Forgetfulness is a refinement of Dr. Ebbinghaus' work **(Fig. 1-3)**.   Reinforcement through repeated education and application is the route to retention. "If you want to know a subject, teach it," is true due to the Ebbinghaus effect!

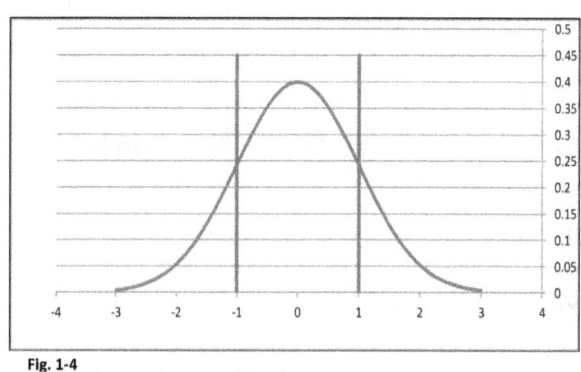

Fig. 1-4

On a normal distribution, +/- sigma is the inflection points on the curve. **SIGMA is real and you can point to it (Fig. 1-4)!**

Carl Friedrich Gauss (1777-1855) developed the idea of a normal curve. But in 1920, Walter Shewhart, working for Western Electric Company at their Hawthorne Works, noted the inflection point in the normal curve and determined that if a process varied by more +/- 3 times that distance from nominal, the process would produce out of print parts. By 1924, Shewhart developed what we now call a Quality Control Chart. I have heard that a symbol and name was needed for the inflection point. The horseshoe Ω omega looked good but the electricians had it. Next was lower case sigma σ.

What we call "Sigma" was, in fact, physically observed (a vision) before mathematics was applied to solve for the location of the inflection point.

When a process yields a normal distribution that just meets the design limits:

At +/- 1 sigma includes 68.27% of the total area under the curve.

At +/- 2 sigma includes, 95.45% of the total area under the curve.

At +/- 3 sigma includes, 99.73% of the total area under the curve.

"%Sigma" is a concept developed for ARMS.

ARMS uses a method to estimate the performance of component suppliers to a project based on the engineer's perception of the supplier's abilities **(Fig. 1-5)**. It is called %Sigma. %σ is not a measured value or an acceptance value. %σ is a judgement from the engineering team that was formed over time through past interactions with the supplier. Whether it is accurate or not is not the point.

It is what the engineers believe and an indicator of where management should apply corrective actions. Verified measurements of supplier performance can occur later as the project moves forward. The use of %σ estimates is intended to inspire preemptive management actions to improve selected areas of the supply base.

ARMS Trumpet Graph (Fig. 1-6): The sample **Median Range** at +/- 1.645σ, (95%) as the sample size is increased 1-35 for units coming

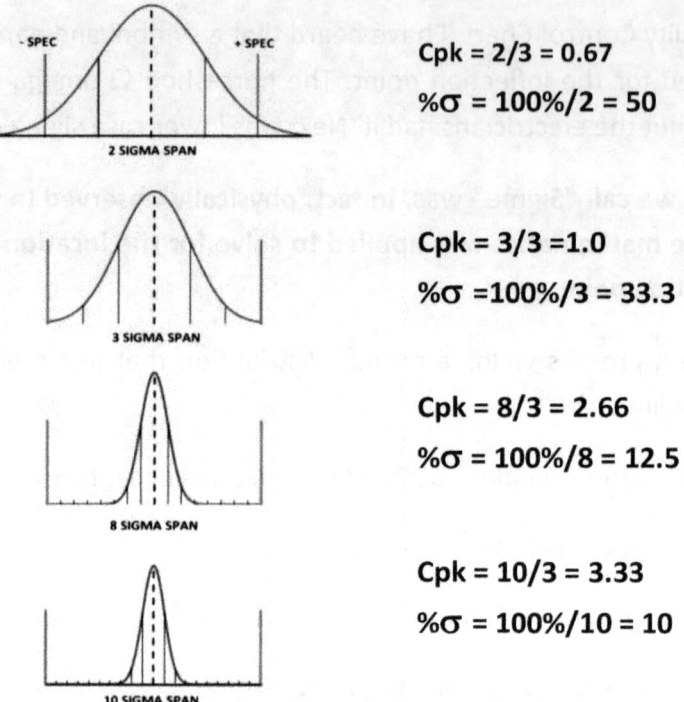

Cpk = 2/3 = 0.67

%σ = 100%/2 = 50

Cpk = 3/3 = 1.0

%σ = 100%/3 = 33.3

Cpk = 8/3 = 2.66

%σ = 100%/8 = 12.5

Cpk = 10/3 = 3.33

%σ = 100%/10 = 10

Fig. 1-5

from sources having a 10, 20, 30, 40 and 50 %σ production capability. We will make use of the plotted +/- 1.95σ points on the Trumpet Graph in our test sample size discussions. %σ was developed for use by ARMS' Workshop engineers.

Forecasts based on the application of sigma need to be approached with caution, in that your involved distribution may not be normal. Distributions may be skewed due to unbalanced forces or may be truncated due to screenings or burn-in activities. It can be useful to plot your data with a linear graph to help determine the distribution conditions present. When you see abnormal distributions, make inquiries. Some physical phenomena may not be continuous and the math may not apply. Limited data samples can also make a normal distribution look skewed or non-normal. Again, use caution when using limited data to make projections.

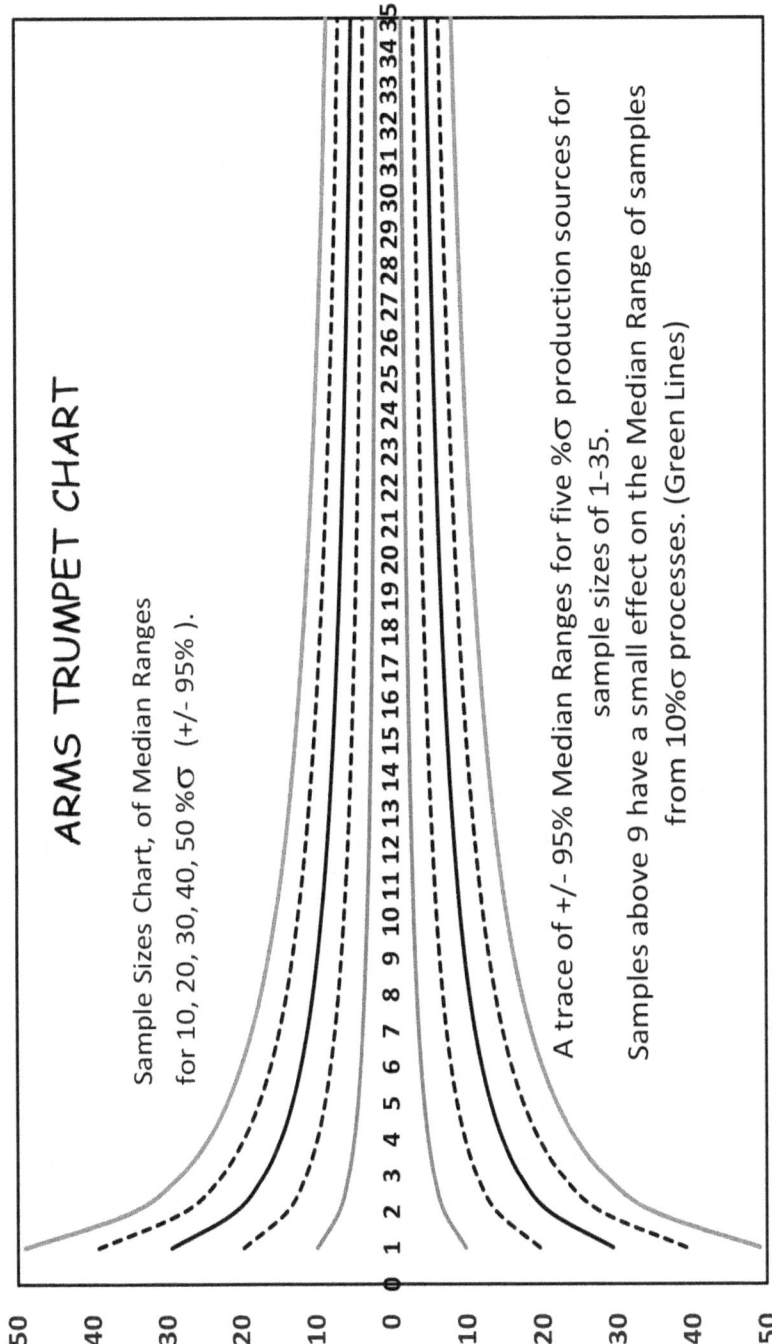

Fig. 1-6

%Sigma was developed to take mystery out of dealing with Cpk values. Cpk values are shown here to satisfy engineers who relate their other work to Cpk values.

%Sigma makes it easier to relate to what the variability represents. Remember for a normal distribution the distance between inflection points is two sigma.

A tightly controlled process might be represented as a 12 %Sigma process (Cpk = 2.6), with all parts to print and most close to nominal **(Fig. 1-5)**.

A process where +/- 3 sigma just covers the range of requirements is a 33% Sigma process (Cpk=1), with 99.73% of output to print.

A more loosely controlled process might be represented as a 50% Sigma process (Cpk=0.67), with 95% of output to print.

SAMPLE SIZE SELECTION

Every project that will be subjected to durability testing is faced with a trade-off on sample size and sample costs. There can be quite a contest between statistical conformation and budgetary limitations. The ARMS' Trumpet Graph **(Fig. 1-6)** is provided to help visualize the effect sample size has on median range. The central horizontal scale is the median by sample size. The left hand vertical scale is for a dimension specification spread over 100% and split into +/- 50% from the nominal median. This demonstrates the effect simple size has on sample ranges from processes with 10, 20, 30, 40 and 50 %sigma variability. As the sample size increases, the 95% median range intervals move closer together and the trace of these points form the trumpet plot.

If processes are not under tight control, larger sample sizes are necessary to reduce the spread and make a more reasonable estimate of the distribution's nominal value. There are two ways to reach a tight distribution of sample dimensions, tighter process capability requirements on producing samples (small %σ) or larger sample sizes. Large sample sizes during early development represent a recurring waste when compared to an investment in the statistical process capability of making samples. Processes yielding values tightly grouped around nominal support utilizing smaller sample sizes to validate the design. Improved accuracy of samples at nominal improves design evaluation and engineering comprehension of results. In the long run, the total cost is less!

Hence, invest in improving process capabilities which will transfer to the next project, rather than waste resources on samples to be destroyed; Investment vs. waste.

A 10 %sigma process with small sample sizes will confirm the design while larger sample sizes simply prove your ability to make parts. 10 %sigma distributions benefit little from sample sizes above 9.

A process with 50 %σ variation requires sample sizes in the 20s or 30s to provide usable results. Remember a 50 %sigma process has 95% of the parts to print. See the 50 %sigma curve, the trace is the outer +/- 2σ line which encloses 95% of the population **(Fig. 1-6)**.

Using fewer samples requires less time, supports your ability to have more design cycle turns for the same money, and leads to a synergistic gain in the rate of learning. In the early stages, it is better to have several redesigns with fewer samples. Larger sample sizes may find random underlying issues in the build process and should be considered later in the design cycle as parts become less expensive. Later, ARMS' Weibull graphs will be employed to forecast into areas where samples cannot be obtained.

There are no blanket answers, but the guideline I recommend is: fewer but more accurate samples in the early stages of design

and to evaluate larger sample sizes in later design cycles.

Caution, samples of one tell you nothing! In essence, frontload the process with rapid design cycle turns where design changes have less cost impact. Then, as parts become cheaper, load the validation facilities. As you initiate production processes, larger sample sizes capture the effect of random occurrence anomalies, which are often due to the new manufacturing processes.

Mil-HDBK 189

The basic formulas for ARMS grew out of working with Mil-HDBK-189 and the Duane postulate as put forward by Dr. Larry Crow. Mil-HDBK-189 is concerned with reliability growth. ARMS has been focused on risk improvement plans, Weibull analysis, and the combined interaction between "Engineering Forecasts" and "measured durability test results" when placed on the same Weibull graph.

From acquired experience, the core formula was changed to an exponential model to improve the fit.

The Duane-Crow AMSAA formula is shown above and the ARMS formula under the curve **(Fig. 1-7)**. Both have been adjusted with a standard unbiased formula to improve small sample size results $1/(n-1)$. The graph plots the results from N=60 units and 5,500 hours for both methods, Duane-Crow AMSAA, above in red and ARMS, below in blue. The upper portion of the Duane-Crow AMSAA result is normally not shown in the text, and a horizontal line is drawn until the result crosses the initial value.

This approach is not necessary with ARMS. Both ARMS and Duane-Crow give the same start and end result. We do not make use of the shape between the start and finish points in either case. ARMS utilizes the exponential approach for clarity. With ARMS, the value of t_o is the total time of testing, λ_o is the failure rate at time zero, λ_i is the incremental failure rate, α is the rate of learning and k is a scal-

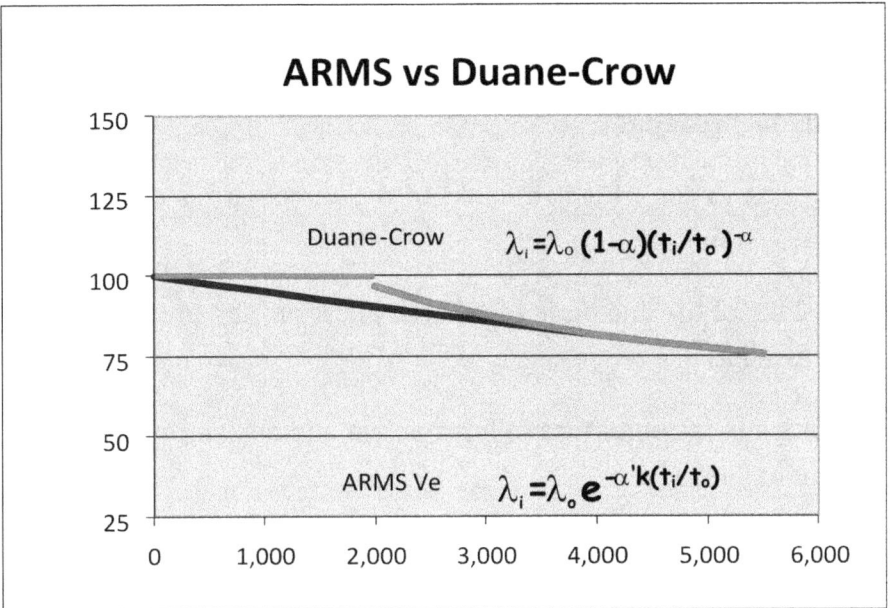

Fig. 1-7

ing factor to allow adjustments to learning due to overstress testing and extended durability testing.

BATH TUB CURVES

To fully appreciate the power of ARMS, it is necessary to have a
deeper understanding of the classical "Bath Tub" curve and how
it is affected by various plotting screens **(Fig. 1-8)**.

(Linear vs Linear) A "Bath Tub" curve depicts <u>non-cumulative</u> failure results during three stages of a product's life.

1) An initial high rate of failure.

2) Lower normal failure rates during useful life.

3) A higher escalating failure rate during wear out.

(Log vs Linear) graph of the same data is plotted to show the effect of using a log scale.

(1) The non-cumulative tub is lifted due to the distortion.

41

(2) The cumulative curve is flattened. The plot curves up at the start, slopes up gently during useful life, then more steeply up during wear out.

('Log-Log vs Log') Median Ranks of the same data is plotted on a Weibull Graph.

The slope of the line during the three phases of the products' life is expressed as β.

(1) β<1 during **premature failures** above and approaching the constant line.

(2) β=1 during **useful life** on the constant random failure line.

(3) β>1 during **wear out,** diverging up from the constant random failure line.

Straight lines with a slope of one on an ARMS' Weibull graph can be thought of as the **"bottoms of bath tub curves."**

When supported by test data, the Weibull lines become demonstrated portions of the bath tub curve and show a bottom to the extent they have been tested. Larger sample sizes will reveal deeper bottoms if they exist.

The data table **(Fig. 1-9)** results in a "Bath Tub" shaped curve when plotted on a Weibull graph **(Fig.1-10)**. The first column is the number of units taken off test, the second column is Failure (1) Suspension (0) and the third is the number of hours on test.

In the **(Fig. 1-10)** ARMS' model, the **(INITIAL)** start value of the forecast was adjusted to an initial 370 / 1000 failures, 37% in one 500-hour year. **(GEN 1)** of the design is forecast to be roughly 38% at 1000 hrs. The Bath Tub curve intersects the **(GEN 1)** design line at about 1000 hours and is parallel to the **(GEN 1)** design line during the useful life period of the bath tub. The bottom of the Bath Tub curve and **(GEN 1)** have a slope β=1. In this example, the forecast start value was calibrated using the measured data. Once calibrated, the **(GEN 1)** design line projects results to be expected at other points. For example: (GEN 1) 23% @ 500 hours, **(Fig. 1-10)**

                              (GEN 2) 13% @500 hours,

                              (GEN 3) 6% @500 hours.

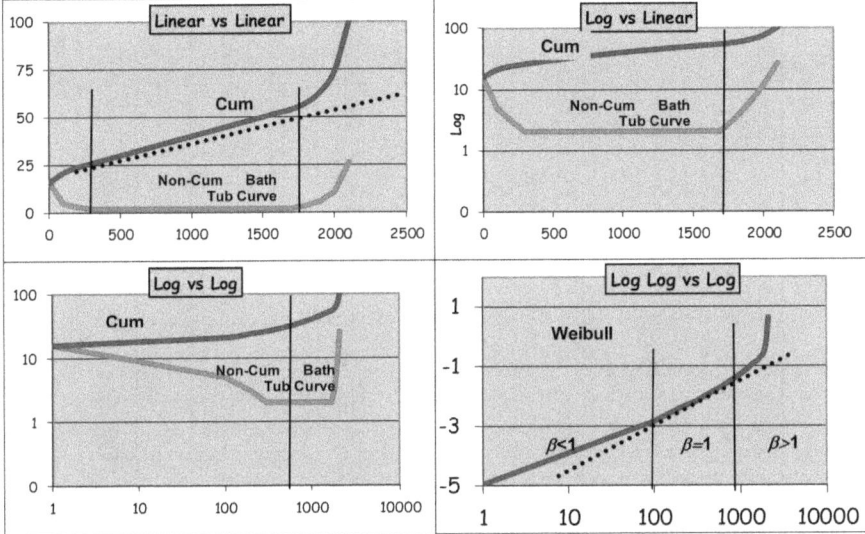

Fig. 1-8

All will fail before the required life goal of 5% @5000 hours.

Often when dealing with data from field applied units, there can be a burst of initial failures. They may be due to application issues or sales' returned units. They may also be from a design that does not take into consideration all factors that must be faced in the field. The result is a bi-model failure set. Bi-model failure sets will be covered in Chapter 10 ARMS' Weibull Analysis discussions.

An ARMS' model requires a starting point

100 units tested to failure		
Bathtub curve		
16	1	1
5	1	100
3	1	200
2	1	300
2	1	400
2	1	500
2	1	600
2	1	700
2	1	800
2	1	900
2	1	1000
2	1	1100
2	1	1200
2	1	1300
2	1	1400
2	1	1500
2	1	1600
2	1	1700
3	1	1800
6	1	1900
11	1	2000
26	1	2100

Fig. 1-9

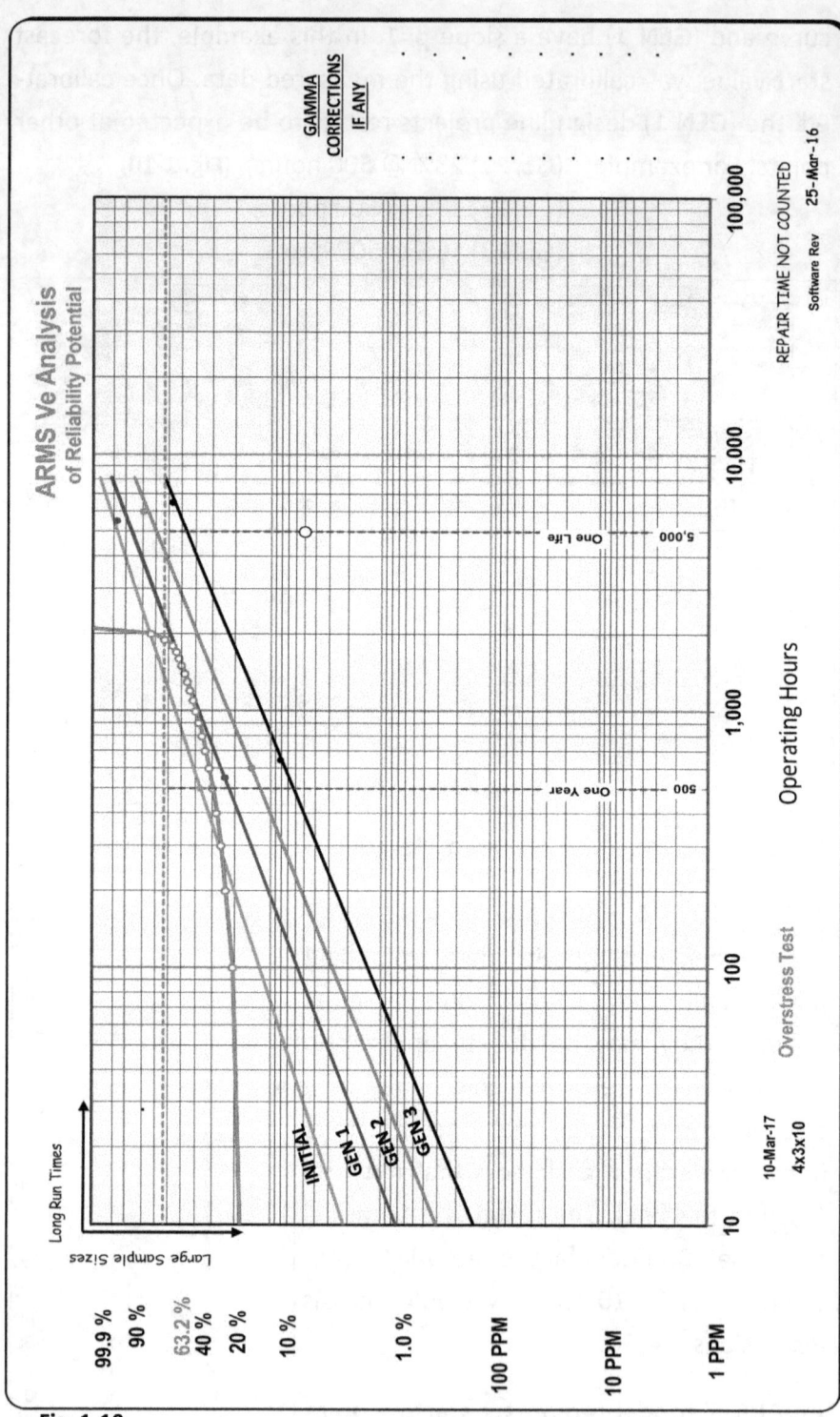

Fig. 1-10

44

failure rate. As we have said, a vision of the design to be modeled exists in the minds of the engineering team and most often in the mind of the lead engineer. The design vision may be from a wide range of possible developments, from a crude sketch in the engineer's mind to an existing design that will be redesigned.

In developing a starting failure rate value, we ask, "If the design, as it exists, were built and 1000 were put into service for one year how many would fail?"

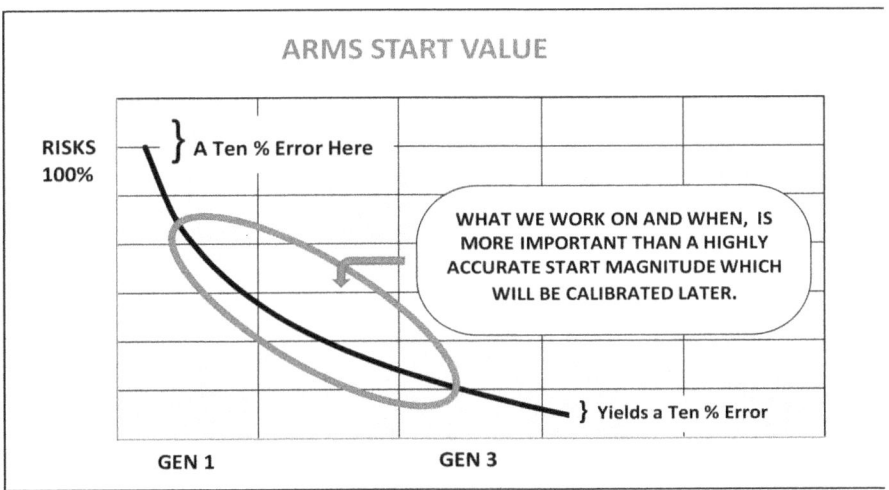

Fig. 1-11

The answer to this question may range from 1000, all of them, to some small number based on earlier designs with field exposure. The answer at this point is not critical, in that we will be adjusting it later as we learn more from our design work and ARMS' analysis. If we expect all would fail, a figure of 500/1000 is applied as a place holder. During initial advanced concept, or innovation design, 500 is often the starting value. The ARMS' START VALUE graph of Failure Rate vs Time depicts the rate of failure improving from your design efforts **(Fig.1-11)**.

Note that a 10% error at the start will result in a 10% error at the end. The end figure is a much smaller absolute value if our design work was successful. Initially in ARMS' analysis, we should not be very concerned with the start or end value. ARMS is about bringing focus and resources to the center of the graph. Namely, what steps are necessary to improve the design? As previously indicated, after we make improvements and are much more familiar with the design, we will be adjusting the "start value" to "calibrate" our ARMS' model utilizing durability test measured results. Again, the engineers are much better at setting the relative % improvement for their work than they would be at setting an accurate absolute start value. After calibration, the ARMS' model will take on much more significance in terms of absolute values. Relative % values prior to calibration are still very valuable for decision making, in that they reflect what the team is thinking and where they think the design risks reside. The engineers are the experts on design, not the accountants. Sorry, guys!

Chapter Two

ARMS' Workshop

and Model

- The largest contributions from ARMS occur during open workshops, when Engineering and Management exchange ideas on what needs to be accomplished, when it needs to be accomplished and what resources may be applied.

- Facilitating Outstanding Leadership / Teamwork between all departments

- Tracking progress can be a messy process, but ARMS keeps us focused on the right tasks and pushes us to take the right actions.

- Engineering Design & Innovation Under Control

ARMS

(Allocated Resource Management System) or as some prefer, (Allocated Risk Management System).

'How to' Instructions:

1) Make a clean work copy of the ARMS' program files on your hard drive, from the source software provided. Your source files must remain "read only." Copies that you have made may be changed by entering data. When asked if a copy should remain read only, type "No." This will permit saving your changes. ARMS requires macros to function and you must adjust your security level to allow ARMS' macros to run. Check with your company IT Administrator if this is a problem.

<u>Important:</u> **Anytime ARMS is moved or transferred to another location or hard drive, all files must FIRST be closed before copying. As with any Excel program, you should not move it if it is open. Using the Windows' file manager, copy all source ARMS' files and paste them into your new location.** If open when moved, Excel will add in pathnames to the old computer location. If this happens, go back to the old location, open ARMS, press UPDATE, then EXIT ARMS and SAVE before copying and moving.

Never use your original source file to input data! ARMS' software is password protected to prevent accidental corruption.

As with all Excel program files, when making changes or adding data:
Do not use "CUT AND PASTE" use **"COPY & PASTE."**
Do not use "DELETE "   use **"CLEAR CONTENTS."**

"Cut and Delete" destroys cell formatting. The original filenames contain revision information. **Prior to use,** change the name of the new file copy using your program nomenclature.

The desired name format is:

For example: "arms-project-name-yymmdd.xlsm"

"arms-lbs-170208.xlsm"

In this example, ARMS Lithium Battery System 2017 Feb 8th. This format style will support sorting by project name and date.

The process of constructing an ARMS' model requires a limited amount of information in advance to help define which model matrix should be selected. A simple block diagram **(Fig. 2-5)** can provide the necessary information. I have had ARMS' Workshops start with as little as a dinner napkin sketch.

ARMS has five matrix frameworks to select from; 4x3x10, 5x4x6, 6x4x5, 6x5x8 and 8x5x6. Our first task will be to fit the new concept into one of these matrices. You will note that the first three matrices, when multiplied out, equal 120. There will be 120 line-items which may be used to describe the full design. With a 4x3x10 matrix there are 4 at a system level, 3 at a sub-system level and 10 each at the lowest level. The last two matrix models multiply out to 240; there are thus 240 opportunities to define your design. It is not necessary to fill in all opportunities, because those left blank do not enter into calculations.

Nearly all designs can be built with one of the first three matrices and you are encouraged to do so. Large systems may require a larger matrix, and they will require a longer time in an ARMS' Workshop to complete. For a team new to ARMS, a day of training preceding the three days to build the first model is normally required. Subsequent ARMS' Workshops each quarter will require at least 2 1/2 days to complete. This, of course, will vary with the complexity of the design. A full Hybrid Electric Vehicle was completed with multiple sets of the 4x3x10 ARMS' matrix. In this hybrid vehicle case, the ARMS' model was also used in reporting results from the suppliers to the program office, and during in-person reports, to the DOT government agency customer. ARMS developed technical progress reports and budget reviews were provided by normal financial oversight. The two reports were made for quarterly oversight reviews.

The open atmosphere resulted in significantly improved communications and confidence between all team members involved. When additional funding was necessary to resolve newly revealed high-risk conditions, it enhanced our ability to level out existing funding and justify additional funding for the program's change in scope.

It is useful to document the team members present during an ARMS' Workshop. An ARMS' log is contained within the ARMS' program for that purpose.

ARMS' Workshops should occur each quarter to keep reliability plans current with new knowledge coming from the development work. The length of time between workshops will depend on the speed of development for the project. Electronics may evolve more rapidly than heavy or large complex products like hybrid vehicles, which contain many sub-systems. Fuel cell propulsion systems required longer development cycles than oxygen sensors or magnetic rheological (MR) shock absorbers. It seems that the longer the time between workshops, the longer the meetings need to be, thanks to the Ebbinghaus effect. With long intervals in between workshops, benefits realized diminish. Two to three months has been the norm for electro-mechanical products, which usually include electronic sub-systems. Effective planning with ARMS results in faster time to production and fewer production changes.

TEAM MEMBERS

If there is a resistance to documenting members' names, it should be overcome. The use of names improves ownership and quality of

Name	FUNCTION	Phone	Enter Date	Date 2	Date 3	Date 4	Date 5	Date 6	Date 7	Date 8
			X							
			X							
			X							

Fig. 2-1

output. If names cannot be utilized, the use of initials may be sufficient. Members will provide stronger support if their name is on the ARMS' model and improvement plan **(Fig.2-1)**.

All members are critical on the team, but three are especially important;

1) The <u>Lead Engineer</u>, where the "Buck" stops.

2) An <u>Engineer</u> assigned to run a laptop computer and,

3) An <u>Engineer</u> capturing Assignable Actions and issues requiring action that come up during the meeting. The position is known as the ARMS' Documenter.

These positions can be combined into one or two engineers, however, guard against overloading the positions. The Lead Engineer needs to make sure the ARMS' Documenter engineer assigned to capturing issues is, in fact, doing a good job.

Risk issues, allocated resources **and the link between an issue and the line items contributing to the risk issue** are the most important outputs of the ARMS' Workshop. An initial ARMS' Workshop typically requires 3-4 days and the ARMS' Documenter, if doing a good job, should be the same person throughout. If a personnel change is made, the laptop goes with the position.

It is recommended that the ARMS' Documenter is an individual who will also become the resident ARMS' leader and have a laptop computer with a minimum of **16 Gs RAM and 256+ Gs of SSD memory.** I use a LENOVO YOGA 910 with 16 Gs of RAM and 512 Gs SSD memory to speed up the execution and provide a very reliable platform. You do not want crashes that are inherent with a mechanical hard drive. ARMS is a very large set of software involving thousands of spreadsheets full of complex calculations. If you do not have the recommended equipment, when you push a button you must be patient and wait for your computer to catch up. 16 Gigs of RAM or more and an SSD (solid state drive) are essential to avoid confusing and frustrating computer delays while conducting an ARMS' Workshop. The YOGA 910 does have a little resolution "quirk," which affects the Excel display. To correct it, set screen resolution to 200% by right mouse clicking on the desktop screen, restart the laptop, reset the resolution to 350% and you are all set. Otherwise, you will not be able to see the type on some programs like Photoshop.

During ARMS' Workshops, a PC projector & screen or a very large HD monitor and a white board with dry erase markers for "chalk talks" will be required. For large rooms, a sound system will be necessary.

To start the ARMS' model, upload a working copy of the ARMS' software files on your hard drive. All files must be in the same folder. Enable macros, then select (load-arms-ve-2017). **ARMS will not function unless you enable Macros.** Do not run ARMS from an external disk or UBS drive to avoid issues caused by Excel linking pathnames into the ARMS' model. Do not run ARMS directly from the source file.

(load-arms-ve-2017) opens the following screen **(Fig.2-2)**. It is im-

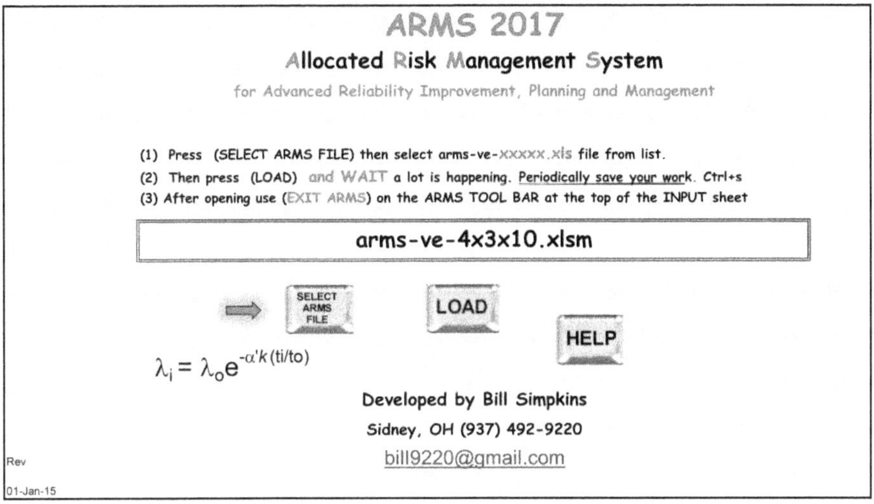

ARMS 2017
Allocated Risk Management System
for Advanced Reliability Improvement, Planning and Management

(1) Press (SELECT ARMS FILE) then select arms-ve-xxxxx.xls file from list.
(2) Then press (LOAD) and WAIT a lot is happening. Periodically save your work. Ctrl+s
(3) After opening use (EXIT ARMS) on the ARMS TOOL BAR at the top of the INPUT sheet

arms-ve-4x3x10.xlsm

SELECT ARMS FILE   LOAD   HELP

$$\lambda_i = \lambda_o e^{-\alpha'k\,(ti/to)}$$

Developed by Bill Simpkins
Sidney, OH (937) 492-9220
bill9220@gmail.com

Rev
01-Jan-15

Fig.2-2

Place all files in the same folder on your hard drive. Do not run directly from DVD disk.

If you are new to ARMS, update your understanding of ARMS, by read the ARMS' Handbook first. Read Instructions sheet below and in each arms-ve-matrix file.

The ARMS' software consists of five arms-ve matrix files, one load-arms-ve file, five arms-ve -tables files:

(1)   arms-ve-4x3x10.xlsm

(2)   arms-ve-5x4x6.xlsm

(3)   arms-ve-6x4x5.xlsm   SAVE with a new name eg. arms-ve-project_yymmdd

(4)   arms-ve-6x5x8.xlsm

(5)   arms-ve-8x5x6.xlsm

(6)   load-arms-ve.xlsm   (START) - Then Select button for an arms-ve model.

(7)   tables-ve-4x3x10.xlsm

(8)   tables-ve-5x4x6.xlsm

(9)   tables-ve-6x4x5.xlsm   NEVER save an OPEN table or change a

(10)   tables-ve-6x5x8.xlsm

(11)   tables-ve-8x5x6.xlsm

Contact Bill Simpkins at bill9220@gmail.com 937 492-9220

portant that you use the "**SELECT ARMS FILE**" button on the left to locate and load one of the ARMS' matrices. The first time you open load-arms-ve-2017, if a file is already listed in the window, press "Select ARMS File" anyway to avoid potential issues caused by Excel linking pathnames from another computer.

After load-arms-ve-2017 has a new file in the window, "**CTRL+S**" and save load-arms-ve-2017. These steps will remove unwanted pathnames. When you want to close your ARMS' model, go to the INPUT tab sheet and press the EXIT button at the top. ARMS will close in an orderly manner. When asked to SAVE, you should do so. During your ARMS' Workshop, you should press **"Ctrl+S"** often to save your work. Future editions of ARMS are planned to utilize ribbons for controls.

There are 12 basic steps to ARMS' Workshops which are iterated as necessary **(Fig. 2-3).**

12 Steps	ARMS PROCESS
(1)	Detailed review of design concept, diagrams and hardware.
(2)	Draw from prior engineering work, if any.
(3)	Select ARMS matrix and build first column from diagrams.
(4)	Assign first year IPT failure rate START VALUE of design as is.
(5)	Allocate risks as percentages. All risks add up to 100%.
(6)	Discuss and set Alpha values, record issues as they arise.
(7)	Discuss and determine any Support Factors, record comments in model.
(8)	Identify opportunities to utilize Engineering problem solving tools.
(9)	Define durability tests, test quantities and Gate Exit requirements.
(10)	Develop, implement & track status of ARMS Assignable Actions.
(11)	Engineering to review ARMS Reports & Forecast with Management.
(12)	Iterate ARMS model in an ARMS Workshop quarterly.

Fig. 2-3

ARMS is a set of software written with Visual Basic in MS Excel. ARMS works with Office 97 forward, but Office 2013 forward is recommended to take advantage of greater speed, smaller file sizes and improved stability. The software contains thousands of spread sheets which are updated frequently.

If you need ARMS' software, please contact: Bill Simpkins, bill9220@gmail.com, for commercial licensed copies and individual shareware.

4x3x10 Matrix			
LEVEL 1 of 4			
	1 of 3	10 ITEMS	
	2 of 3	10 ITEMS	
	3 of 3	10 ITEMS	
LEVEL 2 of 4			
	1 of 3	10 ITEMS	
	2 of 3	10 ITEMS	
	3 of 3	10 ITEMS	
LEVEL 3 of 4			
	1 of 3	10 ITEMS	
	2 of 3	10 ITEMS	
	3 of 3	10 ITEMS	
LEVEL 4 of 4			
	1 of 3	10 ITEMS	
	2 of 3	10 ITEMS	
	3 of 3	10 ITEMS	
	TOTAL	120 ITEMS	
Fig. 2-4	2 of 3	10 ITEMS	
	3 of 3	10 ITEMS	
	TOTAL	120 ITEMS	

Fig. 2-4

Prior to an ARMS' Workshop, one or two engineers will fill in an Excel spread sheet utilizing one of the five basic matrices.

4x3x10 (Yields 120 lines for elements)
4x5x6 (Yields 120 lines for elements)
6x5x4 (Yields 120 lines for elements)
6x5x8 (Yields 240 lines for elements)
8x5x6 (Yields 240 lines for elements)

95% of ARMS' models to date have been built with a 4x3x10, 4x5x6 or 6x5x8 matrix.

With a 4x3x10 matrix, the design is represented by four major divisions, each of which are broken down further into 3 sub-divisions and each of the 3 sub-divisions into 10 parts. It is not necessary that every box be filled in. The matrix merely provides opportunities to enter information where needed. Empty boxes have no effect. However, there must be at least one entry in each of the three levels. Initially, there will be a little confusion while the appropriate matrix and design breakdown is selected and fitted. If a design block diagram exists, it will be helpful in speeding up the process **(Fig. 2-5)**.

With or without ARMS, engineers are expected to make decisions on how to allocate resources during a project. With ARMS, the decisions are visible, reviewable and actionable. The allocations made are the beginning of captured knowledge. Again, throughout the workshop one engineering team member is assigned to capture is-

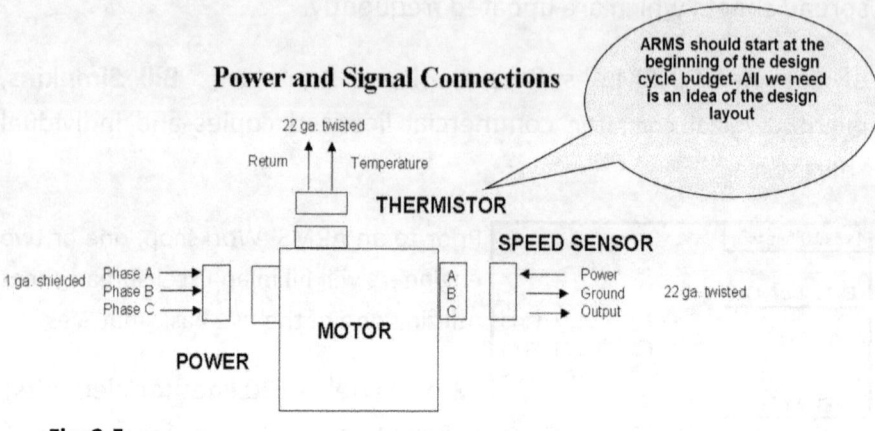

Fig. 2-5

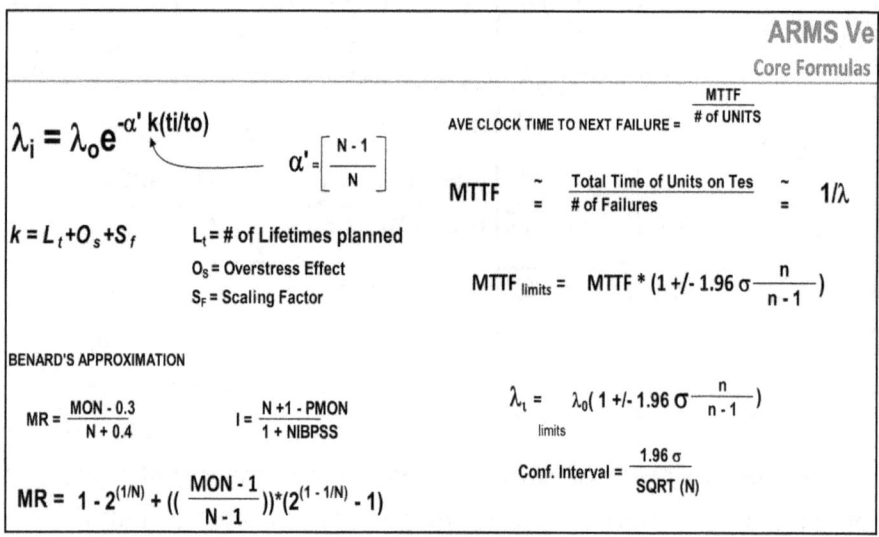

Fig. 2-6

sues and ideas that cross the conference room table. The ARMS' Documenter position is more than recording, it is the most important ARMS' position and needs an engineer who can recognize the importance of an issue and proposed actions during discussions, then probe the team for a clear assignable work statement. Information captured will be developed into an ARMS' reliability improvement plan.

During the ARMS' Workshop, typically the next day, the full engineering team will review the matrix, develop it, make adjustments as necessary, discuss relative risks and add % risk values for each segment of the design. The total risks being allocated add to 100% at each level. Absolute magnitudes are not known and are not needed at this time. We are capturing the team's present understanding of the relative risks within the design.

The "AS IS" nature of the design needs to be stressed, because we are not discussing the design after improvements. The gain from future improvements will be addressed in future quarterly ARMS' Workshops.

The built-in equations, as shown in "Core Formulas" (Fig. 2-6), are applied by ARMS to provide useful graphic outputs. From studying the graphic outputs and how outputs are effected by input variations, engineers will be able to gain new insight from these methods and make the correct decisions.

Individual engineers are encouraged to make the case for why their section has more of the total risk and why they should be receiving a larger allocation during the budgeting of resources. This is a competitive environment, which needs to be encouraged and built upon. The supposition is that budget allocations will follow, and this is the time to **"get your piece of the pie"** (Fig. 2-7).

As engineers, we need to make sure that our issues requiring resolution are being captured in the log of Assignable Actions and linked to the line numbers they affect. The concept is: the more linkages, the greater the allocated resources.

While it may sound simple, actual open discussion provides the engineer an opportunity to put unresolved issues on the table and discuss them with the team, in order to determine their relative risk and to receive resource allocations. **This is where the power of ARMS resides.**

ARMS Ve Reliability Improvement 18-Oct-16

COMPLETE TEAM SHEET	6x5x8			IPT Start IO	5	20	PROG
	HOURS			ONE YEAR TIME	500	365	ON
	CYCLES			END of LIFE YEARS	5.0	1.5	T...RSTRESS
Project Program Manager - Carl K.				END of LIFE GOAL	0.10%	1.2	T...FETIMES
A.D. Engineer - FRED N. 834-5555	Revision 01-Jan-15			END of LIFE CYCLES	1,825	33	SAMP...LE SIZE

ITEM		D.O.E.-R.E. / TRIZ	FEA / GOLDFIRE INNOV.	RED X / ENG. STUDY	Alpha a	Support Factor	IPT II W/O (SF)	IPT II W (-SF)	Alpha a	Support Factor	IPT II W/O (SF)	IPT II W (-SF)	Alpha a	% New Alloc
							GEN 1							GEN 3
11	GRIDS	20%			1.00		6.46	7.67	1.00		3.56	3.66	1.01	1.82
12	MIX	100%	25%	3.00			1.64	1.89			0.81	0.91		0.438
22	COPPER GRID		40%	4.80			2.50	2.91			1.38	1.38		0.688
32	GRID COATING		25%	3.00			1.63	2.04			0.98	0.98		0.472
42	ALUMINUM GRID		10%	1.20			0.69	0.83			0.39	0.39		0.225
52				0.00										
62	FILMS	35%					12.94	14.55			8.35	8.35		5.61
65	ANODE MIX		20%	4.20			2.35	2.72			1.35	1.35		0.775
73	ANODE FILM		20%	4.20			2.57	2.93			1.52	1.52		0.776
83	CATHODE MIX		40%	8.40			5.42	5.93			3.65	3.65		2.669
93	CATHODE FILM		20%	4.20			2.60	2.96			1.82	1.82		1.385
103				0.00										

Fig. 2-7

Agreed upon risk allocations at each level of the model within the team is a very powerful position to achieve. ARMS' discussions always lead to new solution paths. For a few moments, the full team is focused on each engineer's problem and the planned path to resolution. If there are disagreements on how to resolve a risk problem, this is the time to air the concerns. During the risk allocation process, the total risk is defined as 100% at each of the three levels. The absolute magnitude of the risk, at this point, is not important. We are identifying where the risks reside within the design. **If you know where the risks live and their relative magnitudes, then you can manage solutions.** If you do not have agreement across the team including management, ARMS is how you can reach agreement and get everyone on the same track. Engineers are usually good at identifying risk locations, and discussions aid in identifying the relative magnitudes. ARMS supports Management in how to reduce risk. Later, we will get into determining measured magnitudes and calibrating the ARMS' model. This phase of the workshop typically takes an 8-hour day or more depending on the complexity of the design. The team needs time to discuss the issues. This is not a time to rush through it or get pulled out for other meetings. There is a commitment required. Offsite locations are not required, but are very beneficial.

At this point in the project, the ARMS' Weibull Graph may be interpreted to make an estimate of how many "Design Turns" are expected before achieving a design goal. Most often this is a "bad news" revelation. Do not be discouraged by the forecast, realize that the team is working at its peak capability. During development, there are always breakthroughs in design. With ARMS, we are optimizing engineering actions and support by forecasting a continuum of progress. In real life, knowledge grows in steps and jumps. There are disruptions and discontinuities caused by new revelations. Continuously review your design and the ARMS' risk assessments. If there are a few major items assigned high risks, they are candidates for special engineering tools such as focused bench tests, finite analysis and design of experiments' activities. About every three months, it is necessary to build a new ARMS' model in a workshop starting from current conditions.

The largest contributions from ARMS occur during open workshops, when Engineering and Management exchange ideas on what needs to be accomplished, when it needs to be accomplished and what resources may be applied. Detailed high-risk issues, resource limitations and any unrealistic expectations are on the table at the start. **Tracking progress can be a messy process, but ARMS keeps us focused on the right tasks and pushes us to take the right actions. Remember, ARMS is a perfect teambuilder.**

I have conducted many first-time workshops with teams new to ARMS who claimed they had excellent communications. But then it became quite clear that members of management had never sat with a team and jointly discussed the limitations the engineers were working under. Often I have heard statements like, "Well you never asked for that. We can get that for you." or "Now that you have made it clear, we can provide it when it is timely to do so." I have had engineers say that they never asked for something because they knew it would not be approved, and they were wrong. Asking in an open environment may have helped. Managers are not expected to know everything or be focused on design details, that's why they employed smart engineers. Managers are focused on broader responsibilities, but they need to get closer to the engineering team. ARMS is a non-threatening way to do that. The information flow travels in both directions and a cohesiveness develops that is otherwise very difficult to obtain, unless you are picking up the bar bill. Though less true with teams that have been together for a long time and worked many projects, there are still gains that strengthen bonds and open minds to fresh thoughts. Managers with strong engineering backgrounds have an opportunity to make technical contributions based on their years of experience, without becoming bogged down in the project development details. The big picture is up for review as well as the perceived important details that put the project at risk. I have not yet met a manager who did not acknowledge the contribution ARMS was making. Some, like those of the Lithium Battery development and an Advanced Braking System, said ARMS revealed situations that saved the project from a disaster. Not that it is all roses. I did have one team led by a strong engineer who thought management was interfering and had the team resist participation. After a few sessions, this attitude was turned around and success was obtained.

Continuing with the development of the ARMS' model: The upper portion of the ARMS' model INPUT sheet needs to be filled in **(Fig. 2-8)**. This is normally done by one engineer with the support of an ARMS' leader prior to an ARMS' Workshop. At the start of a new workshop there will be an ARMS' education session to level ARMS' knowledge within the team. Having copies of this ARMS' handbook available during the week preceding the ARMS' Workshop will make the training more meaningful. After ARMS' education, data previously applied will be reviewed by the team and adjusted as necessary. Remember Dr. Ebbinghaus **(Fig. 1-3)**!

(1) Assign a Start Value for the failure rate (INPUT, box G3). This is the initial estimated 1st year failures per thousand in a nominal application. For this example, it was set to 100 IPT (Incidents per thousand).

(2) Next set the process %Sigma (INPUT, box H3) as previously defined. E.g., 33% which is a Cpk of 1, where the +/- 3 Sigma limits match the specification limits set by engineering. This is usually the case when suppliers work to print, but no better. If there is a need for management involvement in correcting %Sigma, this is where it can be documented by entering comments into cell H4 of the ARMS' model. During ARMS' Workshops engineers can set tighter specifications for evaluation samples. They can also set worst case design values to be built and maximum test conditions for evaluation.

(3) Enter the operating time for one year of use in normal application conditions (INPUT, box G4). For general automotive operation, 500 hours is an accepted value for hours a vehicle is in motion. 15,000 miles per year is also an acceptable value for travel. Use your company's or customers' usage values. Values requiring determination need to have "Strawman" values inserted into the ARMS' model. Use the comment feature of Excel to explain what the interim values represent. Use "Ctrl+s" often to save your work.

(4) The one year "distance" in miles, kilometers or other units is set and documented (INPUT, box H4). If the design life is measured in cycles, then cycles may be set in box H4. In that case re-label "distance" to "cycles." Leave the sheet tab labeled "Weibull Distance" as is.

(5) End of Life in years is the expected lifetime of the product in normal use. 10 years is a common expected value. Western

Fig. 2-8

		H		K		N	
IPT Start I0	100 #1	33 #2	PROCESS % SIGMA				
ONE YEAR TIME	500 #3	15,000 #4	ONE YEAR MILES				
END of LIFE YEARS	10.0 #5	1.0 #8	OVERSTRESS FACTOR	1.0	TEST OVERSTRESS	1.0	TEST OVERSTRESS
END of LIFE GOAL	5.0% #7	1.2	TEST IN LIFETIMES	1.3	TEST IN LIFETIMES	1.4	TEST IN LIFETIMES
END of LIFE DISTANCE	150,000 #6	6	SAMPLE SIZE	9	SAMPLE SIZE	12	SAMPLE SIZE
	GEN 1			GEN 2		GEN 3	

Fig. 2-9

IPT Start I0	100	33	PROCESS % SIGMA				
ONE YEAR TIME	500	15,000	ONE YEAR MILES				
END of LIFE YEARS	10.0	1.0	OVERSTRESS FACTOR	1.0	TEST OVERSTRESS	1.0	TEST OVERSTRESS
END of LIFE GOAL	5.0%	1.2	TEST IN LIFETIMES	1.3	TEST IN LIFETIMES	1.4	TEST IN LIFETIMES
END of LIFE DISTANCE	150,000	6	SAMPLE SIZE	9	SAMPLE SIZE	12	SAMPLE SIZE
	GEN 1			GEN 2		GEN 3	

Electric Company used 50 years in the 1950s and later found that products, which actually lasted that long, were technologically outdated long before the end of their useful life. Excessive life became a general problem restraining innovation. A Panel Switch for an individual customer line was half the size of a phone booth. A crossbar 5 switch for 100 customers was the size of a small suitcase and digital switches for hundreds of customers would fit on your finger print. The size required shrunk proportionately once optical lines replaced copper wires and microwave horns. Today, the concept of customers connected by switch centers is replaced by an internet with real-time network packet routing. Where did the Central Office switches go? They were replaced by technology, not component failures. The Telephone company's outdated products strongly contributed to the government decision to break the company up. Again, make your determination for life expectations and document them in the ARMS' model with Excel comments.

(6) End of life distance 150,000 miles, in this example, will be calculated from the other inputs made. The end of life value needs to be reviewed for reasonableness and, If necessary, revisit the values that were set for (one-year) and (one life).

(7) An "End of Life Goal" is a business decision. It may have a customer or marketing department origin and their input or concurrence is required. **Zero is not a goal (Fig.2-9).**

(8) The three red outlined boxes under H, K and N are for each of three design generations planned. Set sample sizes that will be built and subjected to validation testing. Refer to the trumpet graph **(Fig. 1-6)** and the beneficial effect of design iterations. During advanced innovation work it is best to have small builds and fast redesign. **One is not a sample size.** As the design moves closer to a production configuration, samples usually become cheaper and are being made on "production intent"

equipment. In that case increase the sample size to capture less frequent random events. This is thus a shift from design evaluation to a combined design and production evaluation. To prove your ability to produce, you need larger sample sizes. To prove a design, you need more accurate hardware.

One program I worked on was called Centra-Cam. It was a compressor design that had a crankshaft with no rod. The piston had a ring with a sliding disk and a hole for the cam shaft. As the camshaft rotated the disk rotated and the piston was forced up and down. There were many perceived advantages in this arrangement that I will not cover here. During engineering development, a few cases occurred where the piston would lock up. Failure analysis always found the piston to be free and in proper working order. A mystery. Development continued and major capital equipment was procured. I even went to Japan to support buying some machines. Engineering believed the mystery failures were due to poor hardware dimensions in engineering machine shop parts. Eventually the Failure Analyst recognized that all such failures occurred when the position was in a top dead center position. Wow, the light bulb lit. In that position, there are no forces on the piston that would apply force to draw it down or up. System Freon gas pressures tended to cause the piston to move to a bottom dead center position, but most times it would only make it part of the way down before it stopped. If not at the top dead center resting position, the compressor would work perfectly well. The problem became an issue of pure geometry. Experts from several major universities were brought in to find a way to solve the problem. But no such luck, the whole program had to be scrapped. The point is, if you made accurate engineering hardware, prejudice would not creep in and so falsely blame machine shop personnel instead of looking more close-

	H			K		N	
IPT Start I0	100	33	PROCESS % SIGMA				
ONE YEAR TIME	500	15,000	ONE YEAR MILES #9				
END of LIFE YEARS	10.0	1.0	OVERSTRESS FACTOR	1.0	TEST OVERSTRESS	1.0	TEST OVERSTRESS
END of LIFE GOAL	5.0%	1.2	TEST IN LIFETIMES	1.3	TEST IN LIFETIMES	1.4	TEST IN LIFETIMES
END of LIFE DISTANCE	150,000	6	SAMPLE SIZE	9	SAMPLE SIZE	12	SAMPLE SIZE
	GEN 1			GEN 2		GEN 3	

Fig. 2-10

	H			K		N	
IPT Start I0	100	33	PROCESS % SIGMA				
ONE YEAR TIME	500	15,000	ONE YEAR MILES #10				
END of LIFE YEARS	10.0	1.0	OVERSTRESS FACTOR	1.0	TEST OVERSTRESS	1.0	TEST OVERSTRESS
END of LIFE GOAL	5.0%	1.2	TEST IN LIFETIMES	1.3	TEST IN LIFETIMES	1.4	TEST IN LIFETIMES
END of LIFE DISTANCE	150,000	6	SAMPLE SIZE	9	SAMPLE SIZE	12	SAMPLE SIZE
	GEN 1			GEN 2		GEN 3	

Fig. 2-11

HOT ZONE MODULE (HZM)
N-CELL STACK MODULE
FUEL REFORMER
CATHODE AIR PREHEAT EXCHANGER (HEX) H2
INTEGRATED COMPONENT MANIFOLD (ICM)
HZM HOUSING
PLANT SUPPORT MODULE (PSM)
PROCESS AIR MODULE
ANODE RECYCLE SYSTEM
PSM ELECTRICAL DISTRIBUTION
FUEL DELIVERY SYSTEM
SYSTEM CONTROLLER (ECU)
APPLICATION INTERFACE MODULE (AIM)
POWER CONDITIONER
AIR FILTRATION SYSTEM
APU PURGE AIR BLOWER B1
AIM ELECTRICAL DISTRIBUTION
ENCLOSURE
STRUCTURAL HOUSING (BASE)
COVER (TOP)
AIM HOUSING
SENSORS
TEMPERATURE INSTRUMENTATION
PRESSURE INSTRUMENTATION
GAS COMPOSITION INSTRUMENTATION
MASS AIR FLOW INSTRUMENTATION
ELECTRONIC MONITORING INSTRUMENTATION
CONTROLS
SAFETY & DIAGNOSTIC SOFTWARE
SOFC CORE PERFORMANCE SOFTWARE
VEHICLE INTERFACE SOFTWARE

Fig. 2-12

ly at the design. But not to worry, in this project they went on to a back-up design and exceeded all of their objectives.

(9) The impact of overstress testing is a science in and of itself. This text will not define the determination methods. Seek help from your test department to set the effect of overstress. With ARMS, you can "What If" the benefits that would be possible with overstress testing. If 20% more failures are expected from overstress, then use 1.2 for the factor. The range will normally be somewhere between 1 and 1.5 before you encounter failures that are not consistent with normal usage. This is another Fermi type estimate that can only be made from within the team. Overzealous setting of overstress can ruin your test program and ARMS' model. Any engineering model is only as good as the sincerity of the input.

(10) Extended life testing is an excellent way to get the most from your invested sample costs. Whenever possible, test until all units have failed. If failures are catastrophic, save a few before failure for analysis. Failures must be supported by excellent failure analysis and the nature of failures must remain consistent with normally applied product failures.

In this and following examples, the data and indentations have an altered structure and values to protect proprietary information

| HOT ZONE MODULE (HZM) | 55.0% |
| N-CELL STACK MODULE | 100% |

GENERATES 42V POWER WITHIN STACK PERFORMANCE SPECIFICATIONS
MAINTAINS GAS SEAL INTEGRITY AND ELECTRICAL ISOLATION
PROVIDES ELECTRICAL CONNECTIONS TO POWER CONDITIONER
MAINTAINS CELL (PEN) INTEGRITY
PROVIDES STACK MOUNTING TO ICM
PROVIDES ANODE OXIDATION PROTECTION

FUEL REFORMER

PROVIDES CONDITIONED AIR FOR FUEL MIXTURE PREPARATION
PROVIDES FULLY VAPORIZED HOMOGENEOUS AIR/FUEL/ANODE GAS MIXTURE TO REACT
REACTOR/CATALYST PROVIDES REQUIRED REFORMATE QUALITY AND POWER OVER OPER
GPC PROVIDES HEAT SOURCE TO CATHODE AIR PREHEAT (HEX) H2
GPC REDUCES EXHAUST EMISSIONS AT OR BELOW REQUIRED LEVELS
MAINTAINS GAS SEAL INTEGRITY

CATHODE AIR PREHEAT EXCHANGER (HEX) H2

PROVIDES HEAT TRANSFER FROM EXHAUST TO CATHODE AIR
MAINTAINS GAS SEAL INTEGRITY

INTEGRATED COMPONENT MANIFOLD (ICM)

REDUCES TEMPERATURE DIFFERENCE BETWEEN CATHODE AIR AND REFORMATE TO STA
MAINTAINS INTERNAL GAS SEAL INTEGRITY
PROVIDES MOUNTING SUPPORT & SEALING FOR STACK MODULE
PROVIDES MOUNTING SUPPORT & SEALING FOR CATHODE AIR PREHEAT EXCHANGER
PROVIDES MOUNTING SUPPORT& SEALING FOR FUEL REFORMER
EFFICIENTLY TRANSPORT PROCESS GASES

HZM HOUSING

PROVIDES THERMAL INSULATION TO ENVIRONMENT
PROVIDES A THERMAL BARRIER BETWEEN THE HOT AND WARM ZONES
PROVIDE SUPPORT OF ICM COMPONENTS WITHIN APU

| PLANT SUPPORT MODULE (PSM) | 20.0% |
| PROCESS AIR MODULE | 100% |

MAINTAINS REQUIRED AIR FLOW & PRESSURE IN THE AIR VALVE MANIFOLD
OPERATES WITHIN POWER CONSUMPTION REQUIREMENTS
OPERATES WITHIN NOISE EMISSION REQUIREMENTS
PROVIDES REGULATED AIR FLOW TO CIRCUITS
MAINTAINS GAS SEAL INTEGRITY

ANODE RECYCLE SYSTEM

PROVIDES REQUIRED ANODE RECYCLE GAS FLOW TO THE AIR/FUEL VAPORIZER
PROVIDES HEAT TRANSFER FROM ANODE RECYCLE GAS TO CATHODE AIR
MAINTAINS GAS SEAL INTEGRITY
OPERATES WITHIN NOISE EMISSION REQUIREMENTS
OPERATES WITHIN POWER CONSUMPTION REQUIREMENTS
PREVENTS BACKFLOW OF FUEL/AIR VAPOR THOUGH THE RECYCLE PUMP

PSM ELECTRICAL DISTRIBUTION

PROVIDES FOR TRANSMISSION OF LOW LEVEL SIGNALS
PROVIDES POWER AND GND BUS DISTRIBUTION
PROVIDES FOR TRANSMISSION OF THE CAN-BUS SIGNALS
MAINTAINS ELECTRICAL CONNECTIONS

Fig. 2-13

and are for illustrative purposes only.

The first example, **Solid Oxide Fuel Cell (SOFC),** was adjusted by the SOFC team to fit into a 6x5x8 matrix. Using the level ONE, TWO and THREE buttons on the INPUT sheet, the cells were filled in as shown **(Fig. 2-12, 13, 14, 15)**. Initially this will take the team a little time until they become more accustomed to the task. If a design does not fit well with the original matrix selected, choose another from the list of ARMS' matrices.

Highly complex systems will require a larger matrix. Normally, you will want as many lines at the third level as possible. (4x3x<u>10</u> , 6x5x<u>8</u>)

The nomenclature applied to each line needs to be unique.

HOT ZONE MODULE (HZM)	55.0%	
N-CELL STACK MODULE	100%	60%
FUEL REFORMER		20%
CATHODE AIR PREHEAT EXCHANGER (HEX) H2		10%
INTEGRATED COMPONENT MANIFOLD (ICM)		5%
HZM HOUSING		5%
PLANT SUPPORT MODULE (PSM)	20.0%	
PROCESS AIR MODULE	100%	50%
ANODE RECYCLE SYSTEM		30%
PSM ELECTRICAL DISTRIBUTION		15%
FUEL DELIVERY SYSTEM		2%
SYSTEM CONTROLLER (ECU)		3%
APPLICATION INTERFACE MODULE (AIM)	7.0%	
POWER CONDITIONER	100%	40%
AIR FILTRATION SYSTEM		10%
APU PURGE AIR BLOWER B1		20%
AIM ELECTRICAL DISTRIBUTION		30%
ENCLOSURE	5.0%	
STRUCTURAL HOUSING (BASE)	110%	60%
COVER (TOP)		10%
AIM HOUSING		40%
SENSORS	8.0%	
TEMPERATURE INSTRUMENTATION	100%	20%
PRESSURE INSTRUMENTATION		5%
GAS COMPOSITION INSTRUMENTATION		40%
MASS AIR FLOW INSTRUMENTATION		30%
ELECTRONIC MONITORING INSTRUMENTATION		5%
CONTROLS	5.0%	
SAFETY & DIAGNOSTIC SOFTWARE	100%	50%
SOFC CORE PERFORMANCE SOFTWARE		25%
VEHICLE INTERFACE SOFTWARE		25%

Fig. 2-14

Having a uniqueness avoids confusion later when we summarize "Assignable Actions" into an improvement plan and report. If necessary, add brackets with a sub-group descriptor such as (HZM) that can be noted when viewing other areas of the ARMS' model. Place the descriptions in the cell in a left / center pattern as shown. Teams have found It best to use **Capital letters in bold Calibri or ARIAL font** for clarity on projection screens and printouts. Color coding by level as shown in ARMS color examples is also recommended.

The ARMS' model does not include lines, which are not assigned a risk, in calculations. If there are entries at level THREE, then an entry will also be required at TWO and ONE. Note the line numbers to the left of each line, as shown on **(Fig. 2-15)**, are for reference. The reference number values jump when there are hidden lines. The hidden lines become visible as we press TWO and THREE and go down deeper into the display levels.

#	Descriptor							
11	HOT ZONE MODULE (HZM)	55.0%						1.000006667
12	N-CELL STACK MODULE	100%	60%					
13				100%				
14	GENERATES 42V POWER WITHIN STACK PERFORMANCE SPECIFICATIONS				25%	0.45	-40%	
15	MAINTAINS GAS SEAL INTEGRITYAND ELECTRICAL ISOLATION				35%	0.50	-50%	
16	PROVIDES ELECTRICAL CONNECTIONS TO POWER CONDITIONER				5%	0.20		
17	MAINTAINS CELL (PEN) INTEGRITY				20%	0.40	-30%	
18	PROVIDES STACK MOUNTING TO ICM				5%	0.30		
19	PROVIDES ANODE OXIDATION PROTECTION				10%	0.40	-75%	
20	FUEL REFORMER		20%					
21				100%				
22	PROVIDES CONDITIONED AIR FOR FUEL MIXTURE PREPARATION				2%	0.20	-10%	
23	PROVIDES FULLY VAPORIZED HOMOGENEOUS AIR/FUEL/ANODE GAS MIXTURE TO REA				18%	0.35	-20%	
24	REACTOR/CATALYST PROVIDES REQUIRED REFORMATE QUALITY AND POWER OVER O				45%	0.40	-20%	
25	GPC PROVIDES HEAT SOURCE TO CATHODE AIR PREHEAT (HEX) H2				10%	0.20	-10%	
26	GPC REDUCES EXHAUST EMISSIONS AT OR BELOW REQUIRED LEVELS				15%	0.30	-50%	
27	MAINTAINS GAS SEAL INTEGRITY				10%	0.30	-20%	
28	CATHODE AIR PREHEAT EXCHANGER (HEX) H2		10%					
29				100%				
30	PROVIDES HEAT TRANSFER FROM EXHAUST TO CATHODE AIR				30%	0.30	-10%	
31	MAINTAINS GAS SEAL INTEGRITY				70%	0.30	-10%	
32								
33								
34								
35								
36	INTEGRATED COMPONENT MANIFOLD (ICM)		5%					
37				100%				
38	REDUCES TEMPERATURE DIFFERENCE BETWEEN CATHODE AIR AND REFORMATE TO S				10%	0.10	-20%	
39	MAINTAINS INTERNAL GAS SEAL INTEGRITY				11%	0.10	-20%	
40	PROVIDES MOUNTING SUPPORT & SEALING FOR STACK MODULE				14%	0.30	-20%	
41	PROVIDES MOUNTING SUPPORT & SEALING FOR CATHODE AIR PREHEAT EXCHANGER				19%	0.30	-20%	
42	PROVIDES MOUNTING SUPPORT& SEALING FOR FUEL REFORMER				21%	0.30	-20%	
43	EFFICIENTLY TRANSPORT PROCESS GASES				25%	0.35	-40%	
44	HZM HOUSING		5%					
45				100%				
46	PROVIDES THERMAL INSULATION TO ENVIRONMENT				30%	0.40	-50%	
47	PROVIDES A THERMAL BARRIER BETWEEN THE HOT AND WARM ZONES				50%	0.35	-40%	
48	PROVIDE SUPPORT OF ICM COMPONENTS WITHIN APU				20%	0.20	-50%	
49								
50								
51								
52	PLANT SUPPORT MODULE (PSM)	20.0%						

Fig. 2-15

When filling in an ARMS' matrix, there are some decisions the team needs to make. At level ONE, it is recommended that the descriptors be "physical components or modules" of the design.

At level TWO, the descriptors may be "Physicals" or they may be "Functions" of the design. Within a subset, the descriptors must all be of the same nature, "Physical or Functional," and not a mix.

The SOFC example has "Physical" descriptors at level ONE and at level TWO, but "Functional" at level THREE (Fig. 2-15). The requirement to not mix types is not a restriction of ARMS, it is a distinction

69

required for human considerations. In our mind, it is not possible to allocate risks across a mix of "Physical" and "Functional" elements.

This 6x5x8 matrix example has 240 lines which may have design elements applied to them. Not all are shown here. You can see that several lines were not needed and are left blank. Also, some placeholders were left in the ARMS' model (D-3, E-3, ……. H-3). They may be removed, if you desire, using "Clear Contents."

In this example, long and very descriptive line items were utilized as an aid to team members and business types who were still getting up to speed on the design.

Long descriptors can help avoid confusion. Shorter descriptors will work as long as they are not redundant from one section to another. If they are, use the coded descriptions for the section (ICM), (HZM) as discussed earlier to differentiate between sections. Within reason, I prefer longer, clearer descriptions to avoid Dr. Ebbinghouse issues.

All risks of failure start at a total of 100% then they are allocated as percentages to sub-sets of the design by team judgement as deemed reasonable. The absolute magnitude of total risk will be determined later and proportional adjustments made at that time. A point to keep in mind is that the team would be making allocation judgements, with or without ARMS, as they determine what will receive attention.

By assigning line #s to the risks in an ARMS' Workshop, the assignments can be reviewed by the team and project leadership. Confirming the completion of the line number assignments is a major review item. The most difficult steps will be when Assignable Actions require the assignment of multiple model line numbers. Line numbers identify everything that will be improved by the assignable action. **The greatest weakness in making these line number assignments is errors of omission and therefore, potential re-source misallocations. Give this action your full attention.**

ARMS documents and reviews current thinking and how the thinking evolved. It also aids in obtaining necessary resources from management and supports your ability to budget resources. If the project is left unbudgeted, issues of the day will chew-up resources and critical issues will go unresolved. Aspirations of solving a problem by conducting expensive design of experiments can be reviewed, justified or deferred to other means or to a more appropriate time. I have observed on several occasions that a DOE was conducted because somebody decided an experiment should be run for the optics it would provide for a customer. They were trying to look good by running a DOE.

Quite often, it becomes obvious that special builds of sub-component sets, not involving full hardware or team participation, are a more efficient solution path. Budget limitations and cash flow are always a factor in the selection and distribution of engineering resources. ARMS is very useful in making the right decisions and getting engineering and management on the same page.

In this example ARMS' model; 100% of the risks, whatever the magnitude might be, were allocated across level ONE design elements as shown **(Fig. 2-14)**. The SOFC Hot Zone Module (HZM) was considered to have the greatest risk of failure. Before arriving at a 55% allocation, there was a great deal of discussion between members of the team. Some felt that the sensors were going to be a significant problem. After reviewing the present status of all areas, the team agreed on the relative risks facing the team. Even the sensor people agreed that while they had a challenge, others faced a greater one.

Valuable cross training and knowledge cross-pollination occurs at these workshops. ARMS' Workshops document what needs to be accomplished and provide an ability to capture knowledge for the future.

Continuing to level TWO, the (HZM) was broken down into five areas as shown **(Fig. 2-14)**. When making these divisions, nothing can be left out. If there are more than five elements, the items need to be collected into five that are fully represented and clearly defined.

Excel comments entered into single cells is the right place to document where clustering has occurred. **Make generous use of the comment capability.** In this example, N-CELL STACK MODULE received 60% of the 55% assigned to (HZM) **(Fig. 2-14)**. If there are some elements of the design that have a low risk assigned, they should be reviewed to determine if this is the time to work on them. When there is a show stopper in the mix, attention and budget need to be focused on prioritizing a solution. This will be covered in chapter 4 (2) in the Sterling thermal engine discussion where work needed to be put into suspension.

Designed experiments may be advisable and resources diverted to support them. During ARMS' Workshops, low hanging fruit easily solved may be identified and put forward even when they are low risk issues. Later, there will be discussions of events that turned up during ARMS' Workshops.

At level THREE we will make our base allocation of risks **(Fig. 2-15)**. It is normal, while filling out level ONE and level TWO, to refer to level THREE descriptions often as a reminder of what lies beneath. **"Ctrl+s" often.**

After the risks are assigned to levels ONE & TWO, assignment of level THREE risks completes the risk allocations. Normally, when discussing level THREE, it is a good idea to proceed horizontally and complete each row item before proceeding to the next row in the set. Determine Alpha learning rates and Support Factors as required. Remember, make generous use of the Excel comment feature to explain why certain values were selected. This is very important to understanding the model under construction and to be able to explain the selections later during Management or Customer reviews if required. Thanks to Dr. Ebbinghaus, you may need the comment explanation as early as next week!

Alpha is the rate of learning expected from actions completed. We initially set the value of Alpha using the ARMS' Alpha Guide **(Fig, 2-16)**. During the Alpha setting phase of the ARMS' model, it is a good idea to have extra hard copies of the guide available for each team member. Team discussions on what Alpha numbers mean will occur and are normal. The discussions help get everyone on the same page.

Alpha is initially set based on four basic considerations:

1) Hardware 2) Application 3) Technology 4) Team Performance.

The initial value of Alpha will be between 0.05 and 0.50. Later it will be necessary to make further adjustments to some of the Alphas and we will address them at the time. The four factors are averaged to a rough value for Alpha, e.g., hardware could be a modification 0.2, a new application 0.3, or understood technology 0.1 and an average value of 0.2 is set.

If there are special actions to increase the learning rate, Alpha may be increased by 0.05 up to 0.2. In this example, Alpha was adjusted to 0.25.

Typically, without special actions, our "rate of learning" (α) peaks and then diminishes the longer we work with an issue. As we proceed from GEN 1 through GEN 3, Alpha values usually decline. If there is an increase in Alpha for a specific item, there will be clearly identifiable reasons which need to be documented with comments in the ARMS' model.

α - (Learning Rate)

ALPHA	Hardware	Application	Technology	Team Performance
0.05	Production	Normal	Cookbooked Solutions	Moderate Alpha values based on specific special actions to increase the learning rate. Robust Engineering, 6σ, FEA, Simulations, Meost, HALT, etc.
0.10	Early Production	Similar	Understood	
0.20	Modification	Less Similar	Developing	
0.30	New Component	New	New to Design Team	
0.40	New Development	New Class for Team	Leading Edge	
0.50	New Dev. & Source	New Class for Industry	Breakthrough	

High values of Alpha normally have an associated high initial failure allocation.

α. Alpha ranges 0.05 to 0.5. Between the line Alpha values are assigned where appropriate. e.g. 0.25 is considered nominal for a well run program. Open ARMS workshop discussions are necessary when setting Alpha. Above normal Engineering actions which increase the rate of learning support an increase in the initial alpha selections. Robust Engineering, FEA, bench tests, Vibration studies, etc.

(SF) Support Factors are based on managerial actions limiting design efforts. e.g.. Overlapping design cycles, evolving specifications, material or supplier changes, decisions to not invest in improving the selected item, test limitations, etc. In some special cases a (SF) of –100% can occur and values of –10% to -30% are not uncommon. Enter comments into input sheet cells.

S F	ENGINEERING SUPPORT RISK	The magnitude of SF is an engineering assessment of support issues
-50%	High	Negative SF's occur when the next Gen of design is committed before testing and learning is completed, when new technology is introduced or solutions come with there own problems.
-40%		
-30%	←	
-20%		Business conditions may not support all possible improvement actions.
-10%		An SF of 100% means no improvement before the next design cycle,
0%	Low	All solutions can be implemented with little or no new problems introduced

Fig. 2-16   74

When assigning Alpha, assume you will be supported by management to do what is deemed necessary. This may be a design of experiments, Shainin red x, finite element analysis or another engineering tool that would increase the rate of learning.

If the team believes needed support for specific activities will not be provided, add a negative Support Factor between -10% and -100%. **Support Factors (SF) are always negative numbers.** An example would be a required test stand not provided at −100%. A test stand delayed might be −30% because some information will be obtained later than needed. Negative Support Factors are a red flag to management. Often, management does not know that the engineering team harbors such thoughts. Frequently, solutions are obtained as soon as the condition is documented. Corrective action by management during an ARMS' meeting is common. If support cannot be obtained, explanations are given that the team can understand. However, in that case, the full learning rate and full −SF remains.

With our SOFC example, level THREE % Risks, Alphas and Support Factors have been entered **(Fig. 2-15)**. While discussing each line

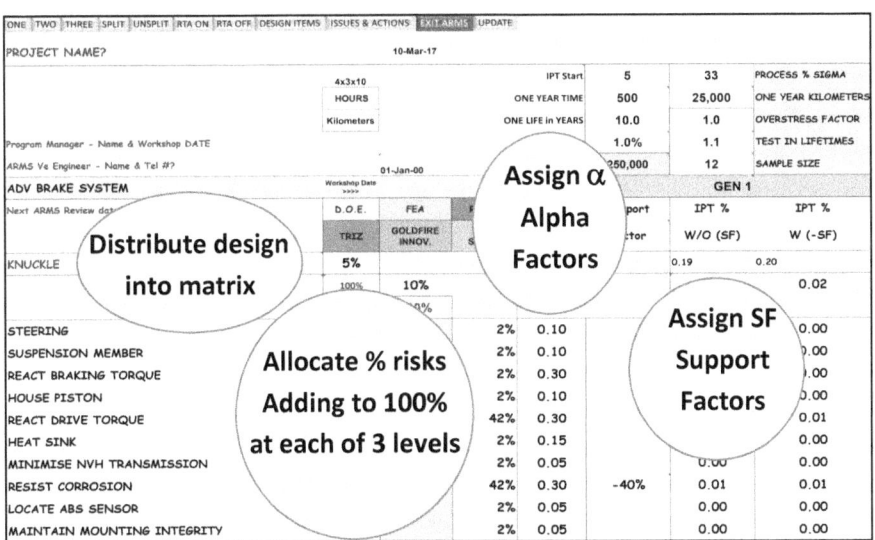

item, the assignment of % Risk, Alpha and Support Factors, and Issues that will require action, are being captured and entered into the model. **Assigning line numbers to actions is critical to the success of an ARMS' model**. Later, these assigned risks will be compiled, and the items will be Pareto ranked by % Risk, for input to a plan of action. In addition to documenting the necessary actions, enter comments into cells explaining why values were set.

Having a risk reduction and design cycle exit gate plan greatly improves the team's ability to deal with unexpected events that occur during all innovation development processes.

Chapter Three

Reliability / Risk Improvement Plans

- All plans shown in this book have been modified to protect confidentialities.

- ARMS supports management by providing an engineering assessment of the challenge to be met.

- Facilitating Outstanding Leadership / Teamwork between all departments

- Engineering Design & Innovation Under Control

The following is an example of less than half of the assigned actions captured during a lithium battery project (Fig. 1-3). Safety Issues are handled under a different process and are not a part of ARMS.

Assignable Actions are linked back to the ARMS' matrix by assigning INPUT level THREE line item numbers to each Assignable Action it affects (Fig.3-1) If more than one item is affected by an action, the line numbers are separated by a comma space convention. If all items are affected, or it is uncertain which are affected, the IPT Index is set to GLOBAL. The assignment of line numbers to actions is a critical operation which will determine the Pareto of the Assignable Actions in the ARMS' report. The ARMS' model sorts the actionable issues into a Pareto list based on the impact each issue is expected to have on the allocated risks entered. This is where the work of the ARMS' Documentation Engineer plays a pivotal role. If diligently entered, the information provides the key output of the ARMS' Workshop effort. Experience has shown that it is critical to support the ARMS' Documentation Engineer in the task to assure important information is not lost during the ARMS' Workshop. As discussed, it is important to review the final risk assignments for errors of omission.

The Pareto listing is the primary source material for developing your Risk Improvement Plan. Typically, the team cannot apply themselves to everything revealed by ARMS, and decisions need to be made as to when issues will actually be improved and completed. There are a multitude of reasons why a straight listing does not become the plan. That is why discernment by the engineering leadership is required. The leader may elect to involve one or two others while forming the Risk Improvement Plan.

ARMS also places design items into a risk Pareto listing. This is useful to review and determine if it is in line with the teams understanding of the design. Agreement insures no one is "sand bagging" a pile of problems. If the design item list does not make sense, then the ARMS' model was not properly constructed and needs to be corrected. An additional column, "Class Utility," is for the use of the team. Numbers, letters or words may be entered that will allow sorting using the column in a manner normal to Excel. Before attempting sorting, make a separate file dated copy of the ARMS' model.

ISSUES	ASSIGNABLE ACTIONS	Line #'s	IPT INDEX	(SF) INDEX	CLASS UTILITY	STATUS
1	ACQUIRE ADHESIVE SPECIFICATION FROM BRAD RELATIVE TO CELL PACKAGING MATERIAL	Global	Global			COMPLETE
53	DETERMINE REASONS FOR EXCESSIVE GASSING DURING FORMATION.	Global	Global			ACTIVE
54	DETERMINE REASONS FOR EXCESSIVE GASSING FOLLOWING DEGAS AND FINAL SEAL	Global	Global			ACTIVE
56	CORRELATE BICELL IMPEDANCE MEASUREMENTS WITH TYPE 1 AND / OR TYPE 2 FADE.	Global	Global			ACTIVE
49	ASSEMBLE ROBUST STUDY ON MIXING AND CASTING OF CATHODE FILM.	85, 86, 87, 88, 89, 95, 96, 97, 98, 99, 100	4.58	-19.2%		PENDING
82	UNDERSTAND CAUSES OF CELL SHORTS - BB PROJECT	178, 25, 44. 119, 129, 140, 141	3.82	-14.6%		ACTIVE
48	ASSEMBLE ROBUST STUDY ON MIXING AND CASTING OF ANODE FILM.	65, 66, 67, 68, 69, 75, 76, 77, 78, 79, 80	3.48	-20.9%		PENDING
13	IDENTIFY ATTRIBUTES THAT CONTRIBUTE TO CAPACITY FADE IN CATHODE ACTIVE MATERIALS	85, 86, 87, 88, 89	2.98	-17.2%		PENDING
38	EVALUATING NEW PACKAGING MATERIALS AND SUPPLIERS TO GET TO 5YR LIFE.	188, 189, 190, 191, 192	2.81	-19.8%		ACTIVE
61	INVESTIGATE DIRECT COATING TO COPPER FOIL	25	2.06	-20.0%		PENDING
59	DETERMINE CAUSE FOR EXCESSIVE GASSING DURING 65 C BAKE PROCESS	89	2.06	-25.0%		ACTIVE
62	UNDERSTAND MIX PROCESS VARIATION DUE TO HUMIDITY	89	2.06	-25.0%		ACTIVE
87	EVALUATE HIGHER TEMPERATURE FORMATION AGAINST LOWER WIMP PROFILE	218, 219	1.97	-12.5%		ACTIVE
14	IDENTIFY ATTRIBUTES THAT CONTRIBUTE TO CAPACITY FADE IN ANODE ACTIVE MATERIALS	65, 66, 67, 68, 69	1.85	-19.8%		PENDING
78	DEVELOP SEALING PARAMETERS FOR POLYPROPYLENE PACKAGING MATERIALS	169, 172	1.56	-20.0%		ACTIVE
83	EVALUATE PECHINEY ON NEW PACKAGING MATERIAL WITH HIGHER ABRASION RESISTANCE	189	1.54	-20.0%		ACTIVE
12	IDENTIFY POTENTIAL SUPPLIERS FOR COATING GRID/FOIL	14, 15, 16, 17, 18, 19	1.36	-18.3%		COMPLETE
63	EVALUATE SOLVENT SYSTEM COMPATIBLE WITH RESIN ON GRID MIX-ADHESION/RESISTANCE	14, 15, 16, 17, 18	1.36	-18.3%		PENDING
81	UNDERSTAND CAUSES FOR STACKED BICELL SHORTS-GB PROJECT	178	1.10			ACTIVE
33	DETERMINE USEFUL LIFE OF CAT DESIGN, WRT LONG TERM PERFORMANCE	167, 169	0.99	-19.6%		COMPLETE
6	SPECIFY THE ATTRIBUTES OF TEMPERATURE RELATIVE TO FORMATION	219	0.99	-10.0%		COMPLETE
43	PERFORM DOE ON ELECTROCHEMICAL IMMOBILIZATION DURING FIRST CHARGE	218	0.99	-15.0%		ACTIVE
44	PERFORM DOE ON FORMATION TEMPERATURE VS. PERFORMANCE	219	0.99	-10.0%		
45	DETERMINE OPTIMUM CONDITIONING PARAMETERS FOR PEAK PERFORMANCE	218	0.99	-15.0%		ACTIVE
86	UNDERSTAND EFFECT OF FORMATION TEMPERATURE ON PERFORMANCE	219	0.99	-10.0%		ACTIVE
88	DEVELOP NEW HIGHER TEMPERATURE NOVRAM OR EQUIVALENT METHOD OF READING AND LOCATING CELL SERIAL NO.	219	0.99	-10.0%		ACTIVE
75	DEVELOP ENVIRONMENTALLY FRIENDLY SURFACE PREPARATION THAT'S STILL EFFECTIVE IN SEALING CELL	169	0.96	-20.0%		ACTIVE
79	EVALUATE ALTERNATIVE PLATING TECHNIQUES ON TERMINAL SURFACE PREP.	169	0.96	-20.0%		ACTIVE
16	DEVELOP AN ELECTRODE THAT YIELD BETTER PERFORMANCE THROUGH THE USE OF HOMOPOLYMER.	197, 198, 199, 200, 202	0.95	-48.8%		PENDING
39	PERFORM ELECTROLYTE VS. PERFORMANCE STUDY FOR ACCEPTABLE PERMEATION.	197, 198, 199, 200, 202	0.95	-48.8%		PENDING
41	PERFORM DOE ON ELECTROLYTE ADDITIVES VS. PERFORMANCE	197, 198, 199, 200, 202	0.95	-48.8%		COMPLETE
52	SPECIFY THE OPTIMUM ACTIVATION ATTRIBUTES (VACUUM, ELYT. QUANTITY, ETC.)	197, 198, 199, 200, 202	0.95	-48.8%		COMPLETE

Fig. 3-1

There are constant evolving timing issues as to when work can be undertaken, how long actions will take and judgements as to how many resources to expend on the effort. Resolving "Support Factors" which inhibit work may itself become an assigned task for resolution. Level loading team members, their skill sets, and the improvement tasks must be resolved and made a part of the Improvement Plan under development. When required skill sets are not found or available within the team, they are only obtainable through management. **ARMS supports management by providing an engineering assessment of the challenge to be met.**

ITEM #	GEN_1 DESIGN ITEMS	START IPT	ALPHA	SF	END IPT	END IPT+SF
89	CATHODE MIX LITHIUM METAL OXIDE	14.0	0.35	-25%	5.11	7.33
25	COPPER GRID CURRENT COLLECTOR	12.8	0.40	-20%	4.04	5.80
189	PACKAGE SHORTING/ABRASION RESISTANCE	10.5	0.35	-20%	3.83	5.17
86	CATHODE MIX PROPYLENE CARBONATE	8.4	0.15		5.45	5.45
218	FORMATION CHARGE PROFILE	7.5	0.30	-15%	3.16	3.81
219	FORMATION TEMPERATURE	7.5	0.30	-10%	3.16	3.59
178	STACK COMPONENT ALIGNMENT	7.5	0.35		2.74	2.74
69	ANODE GRAPHITE	7.0	0.35	-20%	2.55	3.44
190	PACKAGE ELECTROLYTE RESISTANCE	6.3	0.30	-20%	2.66	3.38
169	TERMINAL SURFACE PREPARATION	6.0	0.40	-20%	1.90	2.72
179	STACK BODY WRAP	5.3	0.40	-20%	1.66	2.38
15	GRID MIXCARBON BLACK	4.5	0.35	-20%	1.64	2.21
172	TERMINAL TAPE	4.5	0.30	-20%	1.90	2.42
66	ANODE PROPYLENE CARBONATE	4.2	0.35	-20%	1.53	2.07
76	ANODE FILM POROSITY	4.2	0.35	-30%	1.53	2.33
96	CATHODE FILM POROSITY	4.2	0.20	-30%	2.36	2.91
14	GRID MIX ACRYLIC ADHESIVE	4.0	0.35	-20%	1.46	1.97
36	GRID COAT MOISTURE CONTENT	3.5	0.35	-30%	1.28	1.94
37	GRID COAT ADHESION	3.5	0.30	-30%	1.48	2.08
26	COPPER GRID CLEANLINESS	3.2	0.15		2.08	2.08
44	ALUMINUM GRID	3.2	0.35	-30%	1.17	1.78
202	ELECTROLYTE PURITY	3.2	0.30	-50%	1.33	2.24
75	ANODE FILM COAT WEIGHT	2.8	0.20	-10%	1.57	1.70
80	ANODE HOMOGENEITY	2.8	0.20	-30%	1.57	1.94
87	CATHODE MIX PVDF BINDER	2.8	0.10		2.10	2.10
95	CATHODE FILM COAT WEIGHT	2.8	0.25	-10%	1.36	1.51
100	CATHODE FILM HOMOGENEITY	2.8	0.35	-30%	1.02	1.56
171	TERMINAL DIMENSIONAL SPECIFICATIONS	2.3	0.20		1.26	1.26
177	STACK WELD STRENGTH	2.3	0.10		1.69	1.69
88	CATHODE MIX CARBON BLACK	2.2	0.10		1.68	1.68
191	PACKAGE SEAL STRENGTH	2.1	0.30	-20%	0.89	1.13
78	ANODE CARRIER PROPERTIES	2.1	0.25	-10%	1.02	1.13
98	CATHODE FILM CARRIER PROPERTIES	2.1	0.25	-30%	1.02	1.35
239	SEAL SPECIFICATIONS	1.8	0.30	-35%	0.76	1.12
240	SEAL OPEN CIRCUIT VOLTAGE	1.8	0.30	-50%	0.76	1.28
197	ELECTROLYTE ADDITIVE	1.8	0.20	-50%	1.01	1.41
199	ELECTROLYTE SOLVENT	1.8	0.20	-50%	1.01	1.41
200	ELECTROLYTE SALT	1.8	0.20	-50%	1.01	1.41
67	ANODE PVDF BINDER	1.4	0.15	-20%	0.91	1.01
77	ANODE FILM THICKNESS	1.4	0.20	-10%	0.79	0.85

Fig. 3-2

97	CATHODE FILM THICKNESS	1.4	0.25	-10%	0.68	0.75
117	ANODE ELECT. LAMINATION POROSITY	1.2	0.30		0.51	0.51
127	CATHODE ELECT. CATHODE POROSITY	1.2	0.30		0.51	0.51
68	ANODE CARBON BLACK	1.1	0.15	-20%	0.73	0.81
188	PACKAGE MATERIAL STRENGTH	1.1	0.05		0.91	0.91
192	PACKAGE PERMEATION	1.1	0.30	-20%	0.44	0.56
34	GRID COAT THICKNESS	1.0	0.30	-30%	0.42	0.60
35	GRID COAT WEIGHT	1.0	0.30	-30%	0.42	0.60
38	GRID COAT UNIFORMITY	1.0	0.30	-30%	0.42	0.60
118	ANODE ELECT. FILM TO GRID ADHESION	0.9	0.35	-15%	0.33	0.41
119	ANODE ELECT. FILM CONDUCTIVITY	0.9	0.05		0.78	0.78
128	CATHODE ELECT. FILM TO GRID ADHESION	0.9	0.35	15%	0.33	0.24
129	CATHODE ELECT. FILM CONDUCTIVITY	0.9	0.25		0.44	0.44
45	ALUMINIUM GRID CLEANLINESS	0.8	0.10		0.60	0.60
150	BICELL DIMENSIONAL SPECIFICATIONS	0.8	0.30	-15%	0.34	0.41
167	TERMINAL ALUMINUM	0.8	0.05		0.65	0.65
168	TERMINAL COPPER	0.8	0.05		0.65	0.65
170	TERMINAL CLEANLINESS	0.8	0.20		0.42	0.42
140	SEPERATOR ELECTRICAL RESISTANCE	0.7	0.25	-5%	0.34	0.36
79	ANODE MOISTURE CONTENT	0.7	0.35	-30%	0.26	0.39
99	CATHODE FILM MOISTURE CONTENT	0.7	0.20		0.39	0.39
139	SEPARATOR MECHANICAL STRENGTH	0.6	0.10		0.45	0.45
85	CATHODE MIX ACETONE	0.6	0.10		0.42	0.42
16	GRID MIX ISOPROPYL ALCOHOL	0.5	0.15		0.32	0.32
17	GRID MIX METHYL ETHYL KETONE	0.5	0.15		0.32	0.32
18	GRID MIX DI WATER	0.5	0.15		0.32	0.32
149	BICELL LAMINATION POROSITY	0.5	0.25	-15%	0.24	0.28
228	CELL BAKE TEMPERATURE	0.5	0.20	-90%	0.28	0.48
229	CELL BAKE TIME	0.5	0.20	-90%	0.28	0.48
198	ELECTROLYTE QUANTITY	0.5	0.10		0.34	0.34
238	SEAL VACUUM SPECIFICATIONS	0.4	0.05		0.35	0.35
137	SEPARATOR ADD-ON COAT WEIGHT	0.3	0.25	-10%	0.15	0.16
146	CATHODE/SEPARATOR ADHESION	0.3	0.30	-20%	0.13	0.16
147	ANODE/SEPARATOR ADHESION	0.3	0.30	-20%	0.13	0.16
65	ANODE ACETONE	0.3	0.10		0.21	0.21
136	SEPARATOR POROSITY	0.2	0.25	-10%	0.10	0.11
138	SEPARATOR THICKNESS	0.1	0.25	-20%	0.05	0.06
141	SEPARATOR SHRINKAGE	0.1	0.15		0.06	0.06
151	BICELL PC REMOVAL/DRYING SPECIFICATION	0.1	0.10		0.07	0.07
19						

Fig. 3-3

Engineering leadership retains all of its responsibilities for planning and directing the design program. The output from an ARMS' Workshop clearly identifies what the team believes will reduce risks if addressed and identifies the importance of each item. Engineering leadership works with this valuable input, exercises judgement and makes clarifications and interpretations as needed.

When the Risk Improvement Plan is developed, it is a living document and needs to be periodically reviewed with the engineering team and management members.

In addition to items within the design that need to be addressed, there normally are additional issues dealing with the state of business systems that come out during an ARMS' Workshop.

ARMS is also an educational tool for all involved; the company, the supply base and, I would hope, the customer. ARMS also clarifies the impact "support functions" have on the success or failure of the project.

During development of a Hybrid Electric Vehicle, the Reliability Improvement Plan was considered a prime tool for use in joint customer/company progress reviews and during separate on supplier site, company/supplier meetings. The support from ARMS was greatly appreciated by all parties and created an atmosphere where peripheral opportunities were discussed and new business development considered.

Reliability Improvement Plans are normally treated as confidential material and need to be protected. **All plans shown in this book have been modified to protect confidentialities**.

The full team or team responsibilities are represented in the ARMS' Workshop while the "Assignable Actions" list is being developed. During the Weibull analysis study of forecast and evaluation of test results, not everyone is required. Project leadership will determine which members are needed to support that phase of ARMS and how the Weibull analysis findings will be reported. To understand and interpret Weibull analysis is a specialty. Translations will be needed when making presentations of results.

Chapter Four

Discussions From

ARMS' Applications

- With ARMS, engineers identify where risks reside and their relative magnitudes. We may not know exactly what or how something will fail, but we have a good idea of where the risk of failure lives.

- Facilitating Outstanding Leadership / Teamwork between all departments

- Engineering Design & Innovation Under Control

ARMS supports early changes anticipated from existing knowledge rather than later changes due to test failures. This action is how ARMS breaks the trend shown earlier in the curves on design changes **(Fig. Int 1,2).** ARMS does this by drawing on and optimizing existing knowledge within a design team and by introducing new knowledge which encourages risk reductions and reliability improvements.

With ARMS, the benefit of specialized improvement tools are identified, justified and scheduled before budgets are over committed.

When possible, additional tests should be made with select components to verify worse case limits of the design range and to measure the effect of production variations. There are computer simulations and finite analysis programs that can assist with the variability analysis. Potential applications of finite analysis of component variations and operational limits should be identified during the ARMS' Workshops.

During an ARMS' Workshop, risk allocations, both company and customer risks from product strengths and weaknesses, as conceived in the minds of the engineering team, are considered. This provides tremendous opportunities for managers to assess where limited resources are best applied. It also helps engineers in justifying resource requests. Hard decisions on whether to work on segments of the design or not can be made with the team's knowledgeable buy-in to the decision. Specialized skill sets and equipment necessary for solving high risk issues are determined and plans developed for obtaining them.

The following are discussions from ARMS' Workshops:

1) During a lithium battery development, in an ARMS' meeting, it became obvious that additional chemical engineers were need-

ed if schedules were to be met. The shortage of additional chemical engineering support was slowing the full development team. The delay was causing a waste of budget, time and morale. The manager of the lithium battery facility and the program manager became aware of the problem details through ARMS. The shortage was solved before it became a critical program inhibitor.

2) During an ARMS' Workshop on a Sterling thermal engine development, the ARMS' process identified a major problem that was not being worked on. The engineers had decided to address it later. For the engine to meet efficiency requirements, it needed to operate at $850°$ C or higher. The Sterling engine utilizes an external heat source to heat hydrogen contained in a closed system that pumps hot hydrogen in a loop through the engine, utilizing nickel tubes. At less than $850°C$, the H_2 hydrogen molecules disassociate becoming $-H$ ions which can pass through the nickel grain structure in the tube walls, and be consumed in the external hot zone of the engine requiring constant addition of hydrogen. The H_2 additions increase operating expense and require excessive on-board hydrogen capacity. The team was developing other required components while running all engine tests at lower operating temperatures, or at $850°C$ and supplying make up H_2. Low, unacceptable efficiency was tolerated with the expectation of solving the leakage problem at some later date. They set the problem aside because they did not yet know how to approach the physics required to solve the leakage problem. There was no vision of a route to a solution. It was crystal clear they were developing an inefficient engine that would have no applications. Customer development funds were being spent with no known solution or vision of a solution path. Private comments were even made that the problem may not be solvable. Due to the ARMS' findings, all efforts were diverted and refocused on

achieving a scientific solution to the hydrogen leakage problem first. Without a solution, there would be no need to solve the other issues currently receiving limited financial resources. Solving the hydrogen leakage problem would require a type of engineering expertise that was not available within the current team.

Academic reviews with an external metallurgist suggested exploring Beryllium as a barrier. Beryllium has a tight grain structure. Beryllium should be able to block or sharply reduce the −H ions. However, Beryllium Oxide dust is a highly toxic carcinogen. This requires Beryllium parts to be manufactured in specialized facilities. The primary fabrication of Beryllium components occurs at Los Alamos, New Mexico. After several weeks, meetings with a group of Los Alamos National Labs scientist was arranged. The meeting resulted in a major step toward solving the leakage problem. Therefore, Sterling engine development was able to resume. The complex solution involved gaseous Beryllium plating out onto the ID of nickel tubes, followed by the ID of the tubes being "Ballized" with tungsten carbide balls to cold work the Beryllium, closing any pin holes. The "Ballizing process" draws a ball through the tube and expands the tube size. This approach was something I had previously utilized in 20-foot-long thermistor sensors that wrap around the cowling of all aircraft jet engines. The change in direction was credited with saving a considerable amount of project expense and getting the program back on a productive track. However, the program did stop a few months later for business reasons unrelated to engineering progress.

3) A foreign company, that had been purchased by a large corporation to obtain an advanced brake system design, was struggling to fit into the new corporate structure. The engineering team was falling behind and missing key bench marks in their

development plan. They felt no one was listening to their complaints and moral was low. An ARMS' Workshop was requested by the new parent corporation to be conducted at the off-shore company location. The ARMS' Workshop included a review of the brake design as well as their business systems. Several blind spots were identified in their planned development and it was revealed that the purchase order system for large capital equipment, which required new corporate approvals, was dysfunctional. As a result, engineering purchases were not being processed at headquarters in a timely manner. From the ARMS' Workshop, key contact information was provided and recommended actions were initiated to quickly solve the infrastructure problem. New engineering budgets were set and purchase orders were approved in a timely manner. The basic problem was, "What is the infra-structure system," "Who has the budget" and "Who do you call" when requests are not being processed.

4) During an ARMS' Workshop with a USA manufacturer of medical implant devices, a systemic infrastructure problem was revealed. The medical device, which would be implanted in a patient's back, dispensed medications to the top of the patient's spinal cord. The medical device contained a peristaltic pump, a refillable reservoir, battery, computer and a tube to transport the chemicals to where they were needed. All of the medications were very toxic and corrosive. They required closely controlled handling, application and disposal. The objective of the engineering team was to increase the life of the device from 5 years to 10 years, battery included. Several prototype designs had been put into trial use, prior to involvement with ARMS, as part of an experimental field application program. The ARMS' Workshop revealed that when trial units were returned, they were put into a failure analysis queue behind all "in-production" models returned. Production units required critical reports for legal de-

fense and mandated court review. While waiting for analysis, the engineering units leaked internally and corroded the internal components to the point that no useful information could be obtained. The benefits to be derived from engineering efforts, already expended, were being thrown away while waiting for analysis with all critical knowledge being lost. As a result of the ARMS' Workshop, management was made aware of the infrastructure problem. The Engineering Department, at significant expense, was equipped with a new failure analysis lab with full capability under engineering control. Finally, knowledge was being captured, problems were being solved and progress made. Numerous additional product and process improvements resulted from these ARMS' Workshops.

5) ARMS provided the US Government Department of Transportation program office with an overview of a major hybrid electric vehicle development program involving many related sub-supplier ARMS' projects. A tree of ARMS' projects managed the progress of diverse sub-systems. ARMS coordinated macro scale management adjustments and detailed component level improvements. This activity culminated in a successful program and system level completions. The prototypes developed went on to become the foundation model for production vehicle development.

6) During an initial ARMS' Workshop on advanced electric automotive regenerative brakes, it was revealed that the team lacked a member with sufficient electric motor design and production experience. An expert consultant was identified and able to join the team during ARMS' Workshops and provided the necessary knowledge.

7) During an ARMS' Workshop on a new approach to an electric steering system, it was discovered that the design test require-

DFR Pox Design Items

65	HIGH PRESSURE PUMP	1.650	0.40		0.94	0.94
179	SUPPORT WASHCOAT	1.650	0.50	-30%	0.82	1.07
34	PROVIDE MINIMUM AIR FLOW RATE REQUIRED	1.375	0.40		0.78	0.78
35	PROVIDE MINIMUM AIR PRESSURE REQUIRED	1.375	0.40		0.78	0.78
259	COMMAND INJECTOR BASE PULSE WIDTH	1.155	0.35		0.71	0.71
264	COMMAND AIR CONTROL VALVE POSITION	1.155	0.35		0.71	0.71
265	COMMAND FUEL PRESSURE	1.155	0.35		0.71	0.71
66	PRESSURE REGULATOR	0.825	0.25		0.58	0.58
248	CAN INTERFACE?	0.825	0.30		0.54	0.54
167	NEXTEL WOVEN 312 CLOTH	0.688	0.05		0.64	0.64
258	READ REFORMER MASS AIR FLOW	0.578	0.35		0.35	0.35
262	READ REFORMATE OUTLET TEMPERATURE SENSOR	0.578	0.35		0.35	0.35
263	READ REFORMATE MIX ZONE TEMPERATURE SENSOR	0.578	0.35		0.35	0.35
116	FUEL ATOMIZATION	0.578	0.30		0.38	0.38
180	CONTROL HEAT DISTRIBUTION	0.550	0.30	-50%	0.36	0.46
76	DUCTING FROM PUMP TO REFORMER	0.495	0.15		0.40	0.40
78	DUCTING FROM AIR INTAKE TO PUMP	0.495	0.15		0.40	0.40
86	REFORMATE DELIVERY DUCTING TO EXHAUST	0.495	0.15		0.40	0.40
57	REDUCE NOX TO N2	0.481	0.30		0.32	0.32
187	CONTROL GAS DIFFUSION OF REACTANTS AND PRODUC	0.481	0.20		0.10	0.36
189	INCREASE TURBULENT FLOW	0.481	0.05		0.82	0.45
158	RICH A/F SENSOR	0.440	0.50	-50%	0.22	0.33
159	REFORMATE QUALITY SENSOR	0.440	0.50	-50%	0.22	0.33
126	EXTENDED ELECTRODE	0.385	0.30		0.25	0.25
59	PROVIDES BOOST AIR TO ENGINE INTAKE	0.385	0.25		0.27	0.27
67	PRESSURE ACCUMULATOR	0.330	0.25		0.23	0.23
68	FUEL FILTER	0.330	0.05		0.31	0.31
260	READ RICH A/F SENSOR	0.289	0.35		0.18	0.18
261	READ REFORMATE QUALITY SENSOR	0.289	0.35		0.18	0.18
55	OXIDIZE HC & CO	0.289	0.35		0.20	0.20
58	OXIDIZE PARTICULATES	0.289	0.30		0.19	0.19
44	MAINTAIN MIN. FUEL FLOW FOR REFORMER FUEL PUMP	0.275	0.35		0.17	0.17
181	MINIMIZE PRESSURE DROP	0.275	0.10		0.24	0.24
197	RETAIN CATALYST ASM	0.275	0.20		0.21	0.21
79	COUPLING TO REFORMER AIR INLET TUBE	0.248	0.10		0.22	0.22
198	PROVIDE GAS SEAL	0.206	0.20		0.16	0.16
199	RESIST EROSION	0.206	0.20		0.16	0.16
219	START-UP/SHUT-DOWN ROUTINE	0.206	0.30		0.13	0.13
61	RELEASE STORED SULFUR	0.193	0.35		0.12	0.12
69	TUBING	0.165	0.25		0.12	0.12
65	HIGH PRESSURE PUMP	1.650	0.40		0.94	0.94
179	SUPPORT WASHCOAT	1.650	0.50	-30%	0.82	1.07
34	PROVIDE MINIMUM AIR FLOW RATE REQUIRED	1.375	0.40		0.78	0.78
35	PROVIDE MINIMUM AIR PRESSURE REQUIRED	1.375	0.40		0.78	0.78
259	COMMAND INJECTOR BASE PULSE WIDTH	1.155	0.35		0.71	0.71
264	COMMAND AIR CONTROL VALVE POSITION	1.155	0.35		0.71	0.71
265	COMMAND FUEL PRESSURE	1.155	0.35		0.71	0.71
66	PRESSURE REGULATOR	0.825	0.25		0.58	0.58
248	CAN INTERFACE?	0.825	0.30		0.54	0.54
167	NEXTEL WOVEN 312 CLOTH	0.688	0.05		0.64	0.64
258	READ REFORMER MASS AIR FLOW	0.578	0.35		0.35	0.35
262	READ REFORMATE OUTLET TEMPERATURE SENSOR	0.578	0.35		0.35	0.35
263	READ REFORMATE MIX ZONE TEMPERATURE SENSOR	0.578	0.35		0.35	0.35
116	FUEL ATOMIZATION	0.578	0.30		0.38	0.38
180	CONTROL HEAT DISTRIBUTION	0.550	0.30	-50%	0.36	0.46
76	DUCTING FROM PUMP TO REFORMER	0.495	0.15		0.40	0.40
78	DUCTING FROM AIR INTAKE TO PUMP	0.495	0.15		0.40	0.40
86	REFORMATE DELIVERY DUCTING TO EXHAUST	0.495	0.15		0.40	0.40
57	REDUCE NOX TO N2	0.481	0.30		0.32	0.32
187	CONTROL GAS DIFFUSION OF REACTANTS AND PRODUC	0.481	0.20		0.10	0.36
189	INCREASE TURBULENT FLOW	0.481	0.05		0.82	0.45
158	RICH A/F SENSOR	0.440	0.50	-50%	0.22	0.33
159	REFORMATE QUALITY SENSOR	0.440	0.50	-50%	0.22	0.33
126	EXTENDED ELECTRODE	0.385	0.30		0.25	0.25
59	PROVIDES BOOST AIR TO ENGINE INTAKE	0.385	0.25		0.27	0.27
67	PRESSURE ACCUMULATOR	0.330	0.25		0.23	0.23
68	FUEL FILTER	0.330	0.05		0.31	0.31
260	READ RICH A/F SENSOR	0.289	0.35		0.18	0.18

Fig. 4-1

ISSUES	DFR Pox ASSIGNABLE ACTIONS	Line #'s	IPT INDEX	(SF) INDEX	CLASS UTILITY	STATUS	DUE DATES	LEADER
7	TOTAL VEHICLE SYSTEM INTEGRATION	GLOBAL	GLOBAL			Active	4Q04	Goul/Bon Ham
23	DETERMINE THE MOST VIABLE AIR DELIVERY METHOD	GLOBAL	GLOBAL			Completed	3Q04	Win / Sei
28	DEVELOP SUBSYSTEM SPECIFICATIONS	GLOBAL	GLOBAL			Pending	3Q04	Bon / Silv / Ham
32	REDUCE FUEL ECONOMY PENALTY OF REGENERATION	GLOBAL	GLOBAL			Blocked	4Q04	Goul
34	DEVELOP REFORMER CAPABLE OF USING EXHAUST FOR AIR SUPPLY	GLOBAL	GLOBAL			Active	4Q04	Kir
36	REFINE ENERGY BUDGET FOR THE SYSTEM	GLOBAL	GLOBAL			Active	4Q04	Ham / Goul / Sil
39	DEVELOP NOX ADSORBER AGEING CAPABILITY & SCHEDULE	GLOBAL	GLOBAL					BON
48	DETERMINE SKIN TEMPERATURE MAXIMUM	GLOBAL	GLOBAL					
52	DETERMINE THE EFFECT OF FUEL ADDITIVES ON REFORMER PERFORMANCE	GLOBAL	GLOBAL			Pending		Weis
54	DETERMINE THE ROBUST WINDOW OF OPERATION FOR A/F CONTROL FOR REFORMING	GLOBAL	GLOBAL					DAU / WEIS /
58	DETERMINE THE SENSITIVITY OF THE REFORMER TO FUEL PULSATIONS	GLOBAL	GLOBAL			Active	3Q04	Dau / Sal Lec
73	FIND AN EXTERNAL SOURCE TO SUPPLY REFORMATE CONTROL VALVE	GLOBAL	GLOBAL			Active	4Q04	Ster / Lec
75	DETERMINE DURABILITY OF MASS AIR FLOW SENSOR TO OILS	GLOBAL	GLOBAL					
85	DEVELOP A ROBUST AIR FLOW MEASUREMENT METHOD FOR OIL CONTAMINATED AIR	GLOBAL	GLOBAL			N/A		Li
86	MINIMIZE OIL LEAKAGE INTO AIR FEED STREAM	GLOBAL	GLOBAL			N/A		Win / Sei
108	DETERMINE THE CONDITIONS THAT LEAD TO FUEL PRE-IGNITION	GLOBAL	GLOBAL			Completed	1Q04	Dau / Sal Kir
110	DETERMINE FUEL INJECTOR ABILITY TO SURVIVE IN THE HIGH TEMPERATURE REFORMING ENVIRONMENT	GLOBAL	GLOBAL			Active	3Q04	Ster / Lec
114	DETERMINE THE BEST METHOD TO IGNITE FUEL	GLOBAL	GLOBAL			Pending		Dau
121	DEVELOP A CLOSED-LOOP CONTROL STRATEGY FOR REFORMING	GLOBAL	GLOBAL					ALLST / SAL
126	DEVELOP ROBUST REFORMING FOR FUEL VARIABILITY	GLOBAL	GLOBAL					KIR
138	OPTIMIZE REFORMER PACKAGE FOR DURABILITY AND COST	GLOBAL	GLOBAL					HAM
139	PULL AHEAD ADVANCED MANUFACTURING TASKS FOR ASSEMBLY PROCESS	GLOBAL	GLOBAL			Pending	3Q04	Bon / Sei
141	BEGIN THE ADVANCED SOURCING OF KEY COMPONENTS	GLOBAL	GLOBAL			Pending	3Q04	Bon / Sei
165	IDENTIFY THE CRITICAL DIMENSIONS FOR ENTIRE SYSTEM	GLOBAL	GLOBAL			Pending		Bon / Kir
169	DETERMINE BEST SUBSTRATE FOR REFORMING	GLOBAL	GLOBAL					WEIS
171	DETERMINE THE CATALYST SPACE VELOCITY OPERATING WINDOW FOR PROPER REFORMING	GLOBAL	GLOBAL			Pending		Weis
173	DETERMINE THE CONTRIBUTION OF REFORMER COMPONENTS TO TOTAL PRESSURE DROP	GLOBAL	GLOBAL			Pending		Ham / Nas
177	DETERMINE A METHOD TO ELIMINATE THE REFORMATE QUALITY SENSOR	GLOBAL	GLOBAL			Pending	4Q04	Goul / Ham
178	DETERMINE REQUIRMENTS FOR WASHCOAT DURABILITY	GLOBAL	GLOBAL			Active	3Q04	Weis
179	DETERMINE ROBUST WASHCOAT FORMULATION FOR POx REFORMING	GLOBAL	GLOBAL			Active	38198	Weis
199	EVALUATE ALTERNATIVE PGM'S FOR REFORMING	GLOBAL	GLOBAL					WEIS
200	DETERMINE PGM LOADING FOR DURABILITY	GLOBAL	GLOBAL					
180	OPTIMIZE WASHCOAT TECHNOLOGY FOR COST	187, 188, 189	3.787	-49%				WEIS
69	ALLOCATE PRESSURE DROP REQUIREMENTS TO THE ENTIRE AIR DELIVERY SYSTEM	75, 76, 77, 78, 79, 34,	2.482			Pending		Win / Li/ Ham
77	DETERMINE THE NEED FOR A REFORMATE CHECK VALVE	85, 86	2.341	-96%		Pending	3Q04	Goul / Ham
74	ENLIST OUTSIDE SUPPORT TO DEVELOP A REFORMATE CONTROL VALVE	85	2.247	-100%				
27	DETERMINE THE MAXIMUM INLET AIR SUPPLY TEMPERATURE FOR SYSTEM DURABILITY AND REFORMER PERFORMANCE	116, 77, 65, 55	1.478					STER / LEC
106	DETERMINE OPTIMAL MIXTURE PREPARATION FOR START-UP AND CONTINUOUS REFORMING	116, 77, 65, 55	1.478			Active	4Q04	Dau
57	DEMONSTRATE THE WELDON HIGH PRESSURE SYSTEM	65, 66, 67, 68, 69, 70	1.122					
60	ASSESS THE FUEL SYSTEM DESIGN FOR EVAP AND BARRIER TEST PERFORMANCE	65, 66, 67, 68, 69	1.122					LEC
84	DETERMINE THE NEED FOR THE MIX ZONE TEMPERAURE SENSOR	95	1.021	-50%				
42	DETERMINE THE EFFECT OF COLD/HOT WEATHER TESTING ON SYSTEM OPERATION	54, 55, 56, 57, 58, 59,	0.624					GOUL
66	DETERMINE THE FUNCTIONALITY OF ACV (IAC-TYPE) COMPARED TO ETC-TYPE	75	0.596					
1	IMPACT OF RESIDUAL REFORMATE UPON SHUTDOWN	34	0.591					
49	DETERMINE AUDIBLE NOISE OF REFORMER SYSTEM	34	0.591					
50	DETERMINE THE IMPACT ON THE ENGINE FOR USING TURBOGHARGER FOR AIR SUPPLY	34	0.591					
26	DETERMINE THE NEED FOR INLET AIR TEMPERATURE CONTROL	77	0.485					KIR
67	DETERMINE METHOD TO ELIMINATE EHS	77	0.485					
149	INITIATE ARMS PROCESS FOR RICH A/F AND REFORMATE QUALITY SENSORS	158, 159	0.444	-50%				WAN
43	DEVELOP REFORMATE COMBUSTION STRATEGY FOR VEHICLE COLD	54, 55, 57,	0.393					GOUL /

Fig. 4 - 2

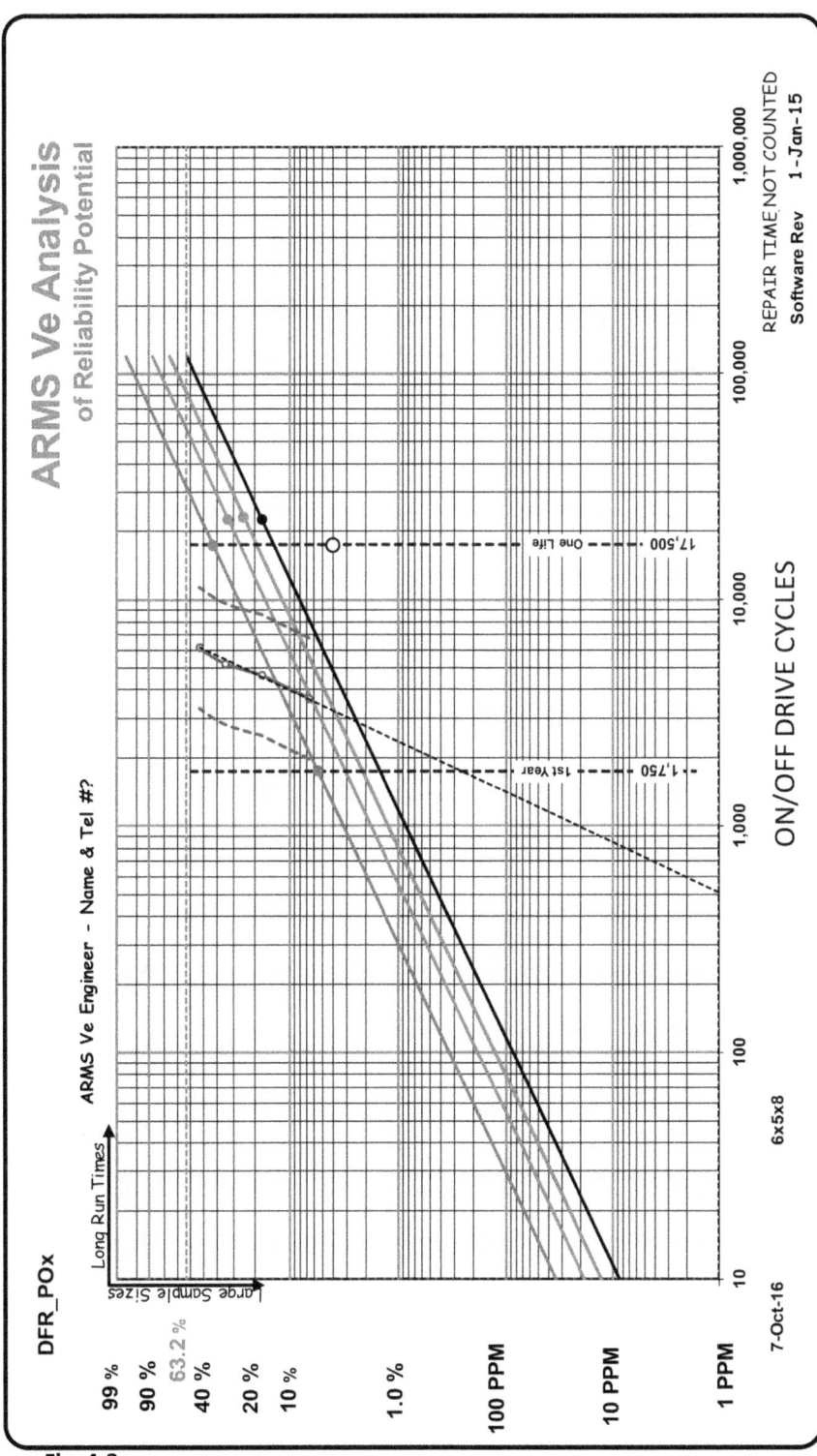

Fig. 4-3

ments did not include a "curb off test." This test simulates parking with the car wheels hard against the curb and turning the

Electro Strut Piston is a basic device consisting of a spool wound with copper wire and leads, over molded to seal, fitted with sealing rings matching the cylinder ID. Applied current provides a magnetic field causing the dampening fluid to thicken.

ITEM	GEN 1 DESIGN ITEMS	START IPT	ALPHA	SF	END IPT	END IPT+SF
19	(3) O-Rings	2.933	0.30	-20%	1.46	1.76
18	Over-Molded Plug	2.346	0.35	-10%	1.04	1.17
51	Inner Socket	1.932	0.35	-30%	0.86	1.18
53	Outer Socket	1.932	0.35	-30%	0.86	0.86
76	Sealing O-Ring	1.932	0.20	-50%	1.21	1.57
17	Electrode Lead	1.760	0.40	-10%	0.70	0.80
16	CoAx Electrode	1.173	0.30		0.58	0.58
20	Magnet Wire & terminations	1.173	0.10		0.93	0.93
21	Final Core Asm Overmold	1.173	0.10		0.93	0.93
28	Rod End O-Ring	1.035	0.30		0.52	0.52
63	Molded Insulator	0.920	0.25		0.52	0.52
52	Inner Flat Terminal	0.828	0.10		0.66	0.66
54	Outer Flat Terminal	0.828	0.10		0.66	0.66
75	Over Molded Connector Asm	0.828	0.20		0.52	0.52
14	Machined Core	0.587	0.15	-10%	0.41	0.43
15	Plastic Insulator - half sections	0.587	0.35	-20%	0.26	0.33
31	Flux Ring	0.414	0.05	-50%	0.37	0.39
26	Finished Piston Rod	0.311	0.10		0.25	0.25
27	Retaining Ring	0.104	0.05	-50%	0.09	0.10
29	Lower End Plate	0.104	0.05	-50%	0.09	0.10
30	Upper End Plate	0.104	0.05	-50%	0.09	0.10

Fig. 4-4

ISSUE	ISSUES & ASSIGNABLE ACTIONS	LINE #'s	IPT INDEX	(SF) INDEX	STATUS	DUE DATES	LEADER
19	RESISTANCE AND VOLTAGE DROP ACROSS TERMINALS AND ELECTRODE AND HARNESS	48	3.92	-25.6%	Complete	20-Jan	M. V.
13	PRESSURE TESTING PLUG IN CORE - CORE OVERMOLD W/NO PLUG O-RINGS, THERMOSET MAT'L	18, 19, 21	3.02	-14.1%	Pending	10-May	M. V.
23	FOLLOW-UP DESIGN REVIEW WITH PACKARD CONNECTION SYSTEMS ENGRG	51, 53, 16	2.74	-23.6%	Pending	21-Apr	M. C.
21	FEA STUDY - PUSH-IN AND PULL-OUT FORCE OF ELECTRODE INTO CONNECTOR TERMINALS	16, 51, 53	2.74	-23.6%	Active	10-Mar	Del
18	WELDED LEAD TO SOCKET TERMINALS - MECH VIBRATION TEST	49	2.49	-25.9%	Pending	28-Apr	M. V.
22	VIBRATION & THERMAL CYCLE TESTING	51, 53	2.15	-30.0%	Pending	28-Apr	M. V.
9	DEVELOP ALTERNATE DESIGN TO ELIMINATE O-RINGS	19, 21	1.71	-17.2%	Active	10-Mar	Igar
10	OBTAIN PROTOTYPE ELECTRODE PLUG ASM'S MOLDED IN PHENOLIC MAT'L	19, 21	1.71	-17.2%	Pending	28-Apr	Igar
11	PRESSURE TESTING PLUG IN CORE - O-RINGS BY THEMSELVES W/NO CORE OVERMOLD	19, 21	1.71	-17.2%	Active	27-Mar	M. V.
12	PRESSURE TESTING PLUG IN CORE - CORE OVERMOLD W/NO PLUG O-RINGS	19, 21	1.71	-17.2%	Active	27-Mar	M. V.
8	DEFINE O-RING PRESENCE ERROR-PROOFING AT CORE SUPPLIER	19	1.47	-20.0%	Active	19-May	Igar
1	TEST SOLDERED JOINT INTEGRITY - MECH VIBRATION W/ THERMAL CYCLING	17	1.06	-10.0%	Pending	28-Apr	M. V.
2	TEST SOLDERED JOINT INTEGRITY - ULTIMATE TENSILE TEST	17	1.06	-10.0%	Pending	24-Mar	M. V.
3	DESIGN STUDY ON ELECTRODE LEAD SOLDERED JOINT	17	1.06	-10.0%	Active	24-Mar	Igar
4	DEFINE SOLDER MAT'L SPECIFICATION	17	1.06	-10.0%	Complete	22-Feb	Igar
24	ROD END ADAPTOR CORROSION PROTECTION - TO BE DEFINED	76	0.72	-50.0%	Active	10-Mar	R. N.
5	TEST FRETTING CORROSION (CONNECTOR END) - MECH VIBRATION W/ THERMAL CYCLING	16	0.59		Pending	24-Mar	M. V.
7	SHIPPING & HANDLING - BARE ELECTRODE - FROM MICRO COAX TO IGAR	16	0.59		Active	19-May	Supplier
20	HIGH PRESURE LEAK TEST ON COAX CONDUCTOR	16	0.59		Complete	10-Jan	W. K.
17	PRESSURE TESTING OF ROD-TO-CORE INTERFACE	26, 28	0.58		Pending	13-May	M. V.
16	SEALING RELIABILITY - SURF FINISH, TIME, TEMP - VERIFIED THRU STRUT DURABILITY TEST	28	0.52		Pending	12-May	M. V.
6	PLASTIC INSULATOR SHELLS DURABILTY - DURABILITY & THERMAL CYCLING TESTS	15	0.33	-20.0%	Pending	24-Mar	M. V.
14	MECH VIBRATION W/ THERMAL CYCLING TO TEST FOR ROBUST WIRE ATTACHMENT	20	0.24		Pending	28-Apr	M. V.
15	IMPACT TEST ON CORE ASM OVERMOLD	21	0.24		Pending	28-Apr	M. V.

Fig. 4-5

wheel to steer free of the curb. Stress on the steering system is at a maximum during this test and far in excess of the stress values expected up to that point. To meet this requirement required a complete redesign. ARMS identified this issue several months and many dollars before it would have been realized had the project proceeded without an ARMS' Workshop.

ARMS' Workshops provide many opportunities to cross-pollinate knowledge between engineers within company, customer and supplier environments. Proprietary information must be protected, but when possible, invite suppliers and customers into your ARMS' meetings. The payback can be tremendous. During one ARMS' Workshop which was attended by a customer, the customer was so impressed that they requested an ARMS' Workshop on the product that would receive the supplier's component, at the customer's location. Customers and suppliers have requested ARMS' Workshops on products not related to the originating company. These actions resulted in significantly closer interactions with a broader range of customer personnel and contracts for business the originating company did not even know existed. Interactions between executive members of the customer, originating company and suppliers normally is as great as a golf outing. This has also been true during ARMS' presentations that precede an ARMS' Workshop.

8) Another example (Fig.4-1,2,3) of an ARMS' model is DFR_Pox, which is a diesel fuel reformer for a Solid Oxide Fuel Cell (SOFC). The ARMS' Weibull of the team's forecast is "calibrated" with test results of ten units and the "Design_Items" are ranked as well as the "Assigned Action Items." For brevity, these are shown without an explanation of the system. Data has been altered from original results for security.

With ARMS, engineers identify where risks reside and their rel-

ative magnitudes. **We may not know exactly what or how something will fail, but we have a good idea of where the risk of failures lives.** This allows us to start reducing the potential of risks at the conception of design work.

9) Energy 5 Electronics is a power electronics bay for a Hybrid Electric Vehicle **(Fig.4-6,7,8,9).** Because this entailed prototype builds during engineering development, there were numerous times the same connectors were plugged and unplugged for a variety of performance evaluations. The number of insertions required during development far exceeded the design capability of the plugs. This is a design fault of the connectors. As a result, con-

ITEM	ENERGY 5 Electronics	Workshop Date >>>> 19-Apr-02			1st GENERATION			
		D.O.E.-R.E.	FEA	RED X	Alpha	Support	IPT	IPT
		TRIZ	GOLDFIRE INNOV.	ENG. STUDY	α	Factor	W/O (SF)	W (-SF)
11	Control Board	20%			1.500		13.7	14.4
44	Power Stage	30%					13.7	14.4
77	Housing/Power Stage Assm	25%					13.7	14.4
110	Field BRIDGE	25%					13.7	14.4
143							13.7	14.4

Fig. 4 - 6

ENERGY 5 Electronics	Workshop Date >>>> 19-Apr-02		
	D.O.E.-R.E.	FEA	RED X
	TRIZ	GOLDFIRE INNOV.	ENG. STUDY
Control Board	20%		
Power Supply	100%	35%	
Digital Core		20%	
Signal Level I-O		25%	
Power Level I-O		20%	
Power Stage	30%		
Heat Sink	100%	10%	
Substrate		20%	
Power Devices		20%	
Interconnects		50%	
Housing/Power Stage Assm	25%		
Lead Frame/Housing	100%	50%	
Components		45%	
Cover		5%	
Field BRIDGE	25%		
Power Devices	100%	75%	
Substrate		25%	

nection failures were the major failure mode. This issue caused a review of connector requirements and a new specification to be placed on future builds. The new connections were significantly better, but still experienced failures. Additional supplier requirements were imposed and restrictions were placed on engineering technicians to

Fig. 4-7

94

#	Item	%	%	%			
11	Control Board	20%					1.500
12	Power Supply	100%	35%				
13			100%				
14	12V Charge Pump				0.40	0.30	
15	5 Volt Regulator				0.30	0.30	
16	Capacitor				0.15	0.05	
17	Inductor				0.15	0.05	-5%
18							
19							
20	Digital Core		20%				
21			100%				
22	Microprocessor HC08				0.50	0.30	
23	Crystal				0.10	0.05	
24	Can Tranceiver				0.40	0.10	
25							
26							
27							
28	Signal Level I-O		25%				
29			100%				
30	Current Sense				0.20	0.10	
31	Temperature Sense(2)				0.10	0.05	
32	Voltage Sense(2)				0.10	0.05	
33	Hall Discrete Input				0.20	0.05	-10%
34	Discrete I-O				0.20	0.05	
35	GATE DRIVERS				0.20	0.15	-20%
36	Power Level I-O		20%				
37			100%				
38	Gate Drive Signals (9)				0.50	0.15	
39	Field Drive Signals (2)				0.30	0.15	
40	Lamp Drive Signal				0.20	0.05	-10%
41							
42							
43							
44	Power Stage	30%					
45	Heat Sink	100%	10%				
46			100%				
47	Transfer Heat				0.35	0.05	
48	Supports Substrate Assm				0.30	0.10	
49	Seals & supports Housing Bottom				0.35	0.05	
50							
51							
52							
53	Substrate		20%				
54			100%				
55	FET Solder joint				0.45	0.25	-10%
56	UTTF conductor				0.30	0.25	
57	adhesive joint				0.05	0.20	-20%
58	printed gate drive resistance				0.10	0.10	
59	Bond Pads				0.10	0.10	
60							
61	Power Devices		20%				
62			100%				
63	Inverter FETs				1.00	0.25	
64							
65							
66							
67							
68							
69	Interconnects		50%				
70			100%				
71	Source to Ground Wirebonds (Long, 45)				0.60	0.35	
72	upper device wirebonds (short, 45)				0.30	0.35	
73	Gate Wirebonds (18, 5 mil)				0.10	0.10	
74							
75							
77	Housing/Power Stage Assm	25%					
78	Lead Frame/Housing	100%	50%				
79			100%				
80	Seals & supports Heatsink				0.60	0.35	
81	Seals and Supports Cover				0.30	0.35	
82	Supports Capacitors				0.10	0.10	
83	Interconnect to Control Board (18)						
84	Supports Control Board						
85	provide power and phase intercconnects						
86	Components		45%				
87			100%				
88	Capacitor				0.40	0.15	-10%
89	Load Dump Diode				0.30	0.20	-10%
90	Sylgard				0.05	0.05	
91	24 pin I/O Header				0.25	0.05	
92							
93							
94	Cover		5%				
95			100%				
96	Label				0.40	0.05	
97	Cover Adhesive				0.40	0.05	
98	Cover				0.20	0.05	
99							
100							

Fig. 4 - 8

schedule all connector disassembly and to replace any connector that reached a specified number of insertions. These actions temporarily resolved the problem during development and resulted in specifying new connectors that would more than meet normal usage insertions in service.

Another problem that led to failures was IGBT power transistors failing under normal load due to overheating during operation. The cause was not immediately clear. This was at a time when power IGBT's were fairly new technology and the engineering community was still learning how to work with them.

At that point, even the internet was still new and had limited information available. After much discussion in an ARMS' Workshop, it appeared that the problem may be the same

ISSUES	ASSIGNABLE ACTIONS	LINE #'s	IPTV INDEX	SF INDEX	CLASS UTILITY	STATUS
1	Bias Supply , Charge Pump etc. to be reviewed for robustness / cost	GLOBAL	GLOBAL			Complete
11	Process improvement for high volume print/ screen adhesive joints	GLOBAL	GLOBAL			Active
12	Process and equipment ID, guidelines for layout of PCB for selective solder	GLOBAL	GLOBAL			Complete
13	Understand what our acceleration factors really are.	GLOBAL	GLOBAL			Active
15	Use best practices in layout of PC board to achieve EMC requirements	GLOBAL	GLOBAL			Active
16	Increase temperature on Crank durability and understand Acceleration factor	GLOBAL	GLOBAL			Pending
17	Redefine thermal Cycling test to increase Acceleration factor of Higher Temperature while testing to failure	GLOBAL	GLOBAL			Pending
18	Obtain Current Surge Data,	GLOBAL	GLOBAL			Active
14	Perform current surge test again to address wire bond length failure (Green Belt	59, 71, 72, 73,	1.29	-2.1%		Active
10	characterize w/b - thicker metallization on die?	71, 73, 83, 85,	0.89	-3.0%		Active
5	Consider / define ASIC for all Gate drives	35, 38, 39,40	0.19	-5.1%		Complete
8	Evaluate solder voiding	55	0.17	-10.0%		Complete
4	Improve Desat performance on Gate drives, make gate drives default to off	38, 39, 40	0.15	-0.8%		Complete
2	Design I/O to meet BCI tests, especially Key On	38, 39	0.13			Complete
6	Review Field Bridge package on ceramic compatibility	39, 48	0.08			Active
3	Check hall signal external power current limiting	33, 34, 35	0.07	-13.7%		Complete

Fig. 4-9

as I encountered with an IGBT high failure rate issue in San Francisco with BART (Bay Area Rapid Transit).

The supplier and BART engineers were stumped by the number of IBGT failures. While all parameters appeared normal, during rush hour failures occurred. It was discovered that the rectifying IGBTs, which switch from positive to negative, were switching when the applied voltage was not quite at zero. The brief transient current spike was high and caused IR^2 heating within the semiconductor. The higher the load current through the IGBT, the greater the spike. For BART, this meant failures at peak operating hours when the rail cars were loaded with customers, not a good combination. Evaluation of the Energy 5 IGBT circuits confirmed the theory and the problem was easily corrected. Large capacitors in the rectification circuit where changing values underload due to internal voltage forces caused foil movement. The solution was to add a high voltage shock as a foil forming process step during production, which physically moved and set the wrapped foils. Electrical short failures of the same power capacitors were also occurring and greater than anticipated. Initially, the team was uncertain of their original ARMS' model estimates for the power capacitors.

Unexpected or unanticipated problems cannot be built into an ARMS' model, they are unknowns. In the case of the power capacitors, it was failure analysis that came to the rescue. During analysis, it was noted that the point of failure was always at the same orientation to the label on the side of the can and appeared to be at the outside diameter of the foil wrap.

A trip to the California capacitor supplier and review of their production process led to discussing the problem with a production oParetor. The oParetor said, "Oh! We do get line rejects for the same reason from time to time due to problems with a cut off step in the automated coil wrap machine that wraps foil on the coils. The cutter can leave a small tab of foil trailing off of the cut edge. That can cause a short circuit when the capacitor flexes during formation under electrical load. The capacitors go through a new forming operation under high electrical load which causes motion in the foils." They had been controlling this coil wrap by having an oParetor make adjustments and sharpening the cutter as needed. Control of the wrap process is an oParetor duty. Recently, they have had a lot of employee turnover at that line position. Action was taken by the supplier to improve the machine and eliminate the possibility of a so-called "hanging chad." The point of discussing this is that things happen that distort expected results, but that does not invalidate the engineering vision present in the ARMS' model.

It is essential to have excellent failure analysis capabilities for the unexpected as well as the anticipated. Also, it is important to allow the Failure Analyst to pursue the solution to supplier locations when needed. Suppliers may not be good at self-examination and reporting. When possible, the analyst should be able to make the call of when to go.

Chapter Five

ARMS

Application Examples

- ARMS documents the team's current understanding of the design and aides in developing a risk improvement plan.

- An ARMS' Workshop is an opportunity for the whole team to learn about the broader challenges the project and company are facing.

- ARMS' reports and ARMS' models are knowledge capture!

- The Engineering Team's thoughts and expectations are being plotted on the same graph as measured results!

- Open Team Interactions, Knowledge Capture, a Pareto of Risk Items, and a Pareto of Assignable Actions which are inputs to forming a Project Plan and Exit Gate strategy.

ARMS is about the interaction between team members during an ARMS' Workshop. The exchange of experience and knowledge incipient to an ARMS' Workshop exceeds all other methods that attempt to draw on team assets I have encountered. The living, dynamic nature of ARMS makes it very difficult to convey the strength of ARMS with a printed example. A review of ARMS' INPUTS and OUTPUTS does not reveal the deep strength and contribution ARMS makes during the workshop. With these descriptive limitations in mind, the following is an application example.

LITHIUM BATTERY SYSTEMS (LBS)

ARMS normally deals with information sensitive to a company, and so this limits what can be published and discussed. I have selected an ARMS' example from a series of workshops on a project that did not go into production for business reasons not connected to the product's design, development or costs. The ARMS' Workshops were conducted on the design and development of a family of lithium batteries. The lithium battery team was part of a much larger lead-acid battery division. The entire battery division was spun off to a foreign entity by the parent corporation during a downsizing action. This ARMS' example involves the product development of individual lithium battery cells, which were later to be combined into a series of battery pack configurations. All items of propriety security concern have been modified or removed to protect company information.

Lithium Battery Systems illustrated in this ARMS' Handbook cannot and should not be used to imply anything about the performance of existing applications, battery construction or design. Application types fall into one of several classes, depending on the requirements of an application. Applications such as laptops, watches, cameras, automobiles, lighting and so on have very different requirements for voltage, current, power, storage

LITHIUM CELL

GRIDS
FILMS
BICELL
PACKAGED CELLS
FORMED CELL
*

Fig. 5-1

LITHIUM CELL

ITEM	
11	GRIDS
12	MIX
22	COPPER GRID
32	GRID COATING
42	ALUMINUM GRID
52	
62	FILMS
63	ANODE MIX
73	ANODE FILM
83	CATHODE MIX
93	CATHODE FILM
103	
113	BICELL
114	ANODE ELECTRODE
124	CATHODE ELECTRODE
134	SEPARATOR
144	BICELL LAMINATION
154	
164	PACKAGED CELLS
165	TERMINALS
175	STACK ASSEMBLY
185	PACKAGING
195	ELECTROLYTE
205	
215	FORMED CELL
216	FORMATION
226	BAKE
236	DEGAS & FINAL SEAL
246	*
256	*

Fig. 5-2

life, operating and storage environment, discharge rate, depth of discharge demand, lifetime charge cycles, rates of charge and several additional requirements. None of these specifics will be addressed in this book. This LBS application example is for a medium to high power, deep discharge application operating within a modest temperature range environment. The application discussed is not an automotive or computer related system.

Each Individual cell is contained within a soft pouch structure and several cell packs are combined to achieve the required voltages and currents. Tests are conducted on special test rigs that simulate pack application conditions, yet permit handling and monitoring the cells individually. In this ARMS' example, "cell" refers to a single battery pouch.

Safety of Lithium Batteries is a very large topic of concern, and issues pertaining to safety were addressed in a separate forum.

GRIDS		
	MIX	
ACRYLIC ADHESIVE CARBON BLACK ISOPROPYL ALCOHOL METHYL ETHYL KETONE DI WATER		
	COPPER GRID	
COPPER CURRENT COLLECTOR CLEANLINESS		
	GRID COATING	
THICKNESS COAT WEIGHT MOISTURE CONTENT ADHESION UNIFORMITY		
	ALUMINUM GRID	
ALUMINUM GRID CLEANLINESS		

Fig. 5-3

The first ARMS' Workshop activity is to "fit" the LBS Cell design, to the extent it is initially understood, into an ARMS' matrix. As previously discussed, the ARMS' matrices are 4x3x10, 5x4x6, 6x4x5, 6x5x8 and 8x5x6. During this ARMS' Workshop, the team settled on a 6x5x8 matrix. At the Lithium Battery System ARMS' Workshop, the team felt they would need a large model that would accommodate individual LBS Cell features and package considerations and still have adequate opportunities to describe conditions at the third level of the model where α and **SF** are applied.

Level_ONE **(Fig. 5-1)** assignments were, GRIDS, FILMS, BICELL, PACKAGED CELLS and FORMED CELLS. The sixth item of level ONE was not needed and therefore left blank. There was a great deal of discussion, during these selections, that revealed some of the concepts team members had in the back of their minds for future action. In some cases, those

		LITHIUM CELL	Workshop Date >>>>
			D.O.E.-R.E.
			TRIZ
ITEM			
11	GRIDS		20%
62	FILMS		35%
113	BICELL		5%
164	PACKAGED CELLS		30%
215	FORMED CELL		10%
266	*		

Fig. 5-4

LITHIUM CELL	Workshop Date >>>>		
	D.O.E.-R.E.	FEA	RED X
LEVEL TWO	TRIZ	GOLDFIRE INNOV.	ENG. STUDY
GRIDS	20%		
MIX	100%	25%	
		100%	
ACRYLIC ADHESIVE			40.0%
CARBON BLACK			45.0%
ISOPROPYL ALCOHOL			5.0%
METHYL ETHYL KETONE			5.0%
DI WATER			5.0%
COPPER GRID		40%	
		100%	
COPPER CURRENT COLLECTOR			80.0%
CLEANLINESS			20.0%
GRID COATING		25%	
		100%	
THICKNESS			10.0%
COAT WEIGHT			10.0%
MOISTURE CONTENT			35.0%
ADHESION			35.0%
UNIFORMITY			10.0%
ALUMINUM GRID		10%	
		100%	
ALUMINUM GRID			80.0%
CLEANLINESS			20.0%

Fig. 5-5

thoughts were brought forward into the current design phase.

Each level ONE design element was broken down into level TWO items (Fig.5-2).

Lines not needed were left blank and have no effect on the calculations. Again, there was a great deal of discussion among team members and an opportunity to capture risk issues deemed important by the team.

Ctrl+s" to save your work frequently.

Remember, the LBS Engineer assigned to capture and record assignable actions is also known as the ARMS' Documenter, a key position on the team.

The ARMS' Documenter responsibility is assigned to one of the top engineers with the right

11	GRIDS	20%			1.00	
12	MIX	100%	25%			
13				100%		
14	ACRYLIC ADHESIVE			40.0%	0.35	-20%
15	CARBON BLACK			45.0%	0.35	-20%
16	ISOPROPYL ALCOHOL			5.0%	0.05	
17	METHYL ETHYL KETONE			5.0%	0.05	
18	DI WATER			5.0%	0.05	
19						
20						
21						
22	COPPER GRID		40%			
23				100%		
24						
25	COPPER CURRENT COLLECTOR			80.0%	0.40	-20%
26	CLEANLINESS			20.0%	0.05	
27						
28						
29						
30						
31						
32	GRID COATING		25%			
33				100%		
34	THICKNESS			10.0%	0.25	-30%
35	COAT WEIGHT			10.0%	0.25	-30%
36	MOISTURE CONTENT			35.0%	0.30	-30%
37	ADHESION			35.0%	0.20	-30%
38	UNIFORMITY			10.0%	0.20	-30%
39						
40						
41						
42	ALUMINUM GRID		10%			
43				100%		
44	ALUMINUM GRID			80.0%	0.35	-30%
45	CLEANLINESS			20.0%	0.05	
46						

Fig. 5-6

FORMED CELL	10%	
FORMATION	75%	
FORMATION CHARGE PROFILE		50%
FORMATION TEMPERATURE		50%
BAKE	5%	
CELL BAKE TEMPERATURE		50%
CELL BAKE TIME		50%
DEGAS & FINAL SEAL	20%	
SEAL VACUUM SPECIFICATIONS		10%
SEAL SPECIFICATIONS		45%
SEAL OPEN CIRCUIT VOLTAGE		45%

Fig. 5-7

mindset and ability to capture knowledge and the important Assignable Actions. Some engineering groups have actually assigned two ARMS' Documenters to assist one another and assure all information was captured. In this case study, note that the last level one group, "FORMED CELLS," is divided into manufacturing processes (Fig.5-7)!

As you can imagine, there was a great deal of discussion over this decision. Processes are not a problem for the ARMS' model as long as we do not mix processes and design elements in the same subgroup. If processes and design elements are mixed in one sub-group, it is tough to keep the allocations clearly separated in our minds.

The LBS team felt that the manufacturing processes were an important part of lithium cell design considerations and resultant reliability. The chemical processes contained high risk elements that required inclusion into the LBS ARMS' model. Adequate moisture control was known to be important, however, the extent of controlling moisture was a challenge learned through experience during the LBS Cell development program.

Including manufacturing processes drove home the necessity of including manufacturing engineering as a full-time member of the ARMS' Workshop team. Discussions after the manufacturing engineering members were onboard allowed the m.e.s to provide critical input to the design team and identify previously unknown risks requiring correction that would not have been discovered otherwise until late in the design and development effort. This action to include key manufacturing processes in the model turned out to be a valuable and critical contribution.

At level THREE (Fig.5-6), the design elements were reduced further. The assignment of design elements to an ARMS' matrix may be made by one or two engineers prior to the workshop and then be reviewed and corrected at the start of the ARMS' Workshop. If elements are assigned prior to the workshop, it should be by or reviewed with the ARMS' Documentation Engineer prior to starting the initial ARMS' Workshop. After the design elements are assigned to an ARMS' matrix and agreed upon by the team, we are

α - (Learning Rate)	Alpha & SF Rational			
ALPHA	Hardware	Application	Technology	Team Performance
0.05	Production	Normal	Cookbooked Solutions	Moderate Alpha values based on specific special actions to increase the learning rate. Robust Engineering, 6σ , FEA, Simulations, Meost, HALT, etc.
0.10	Early Production	Similar	Understood	
0.20	Modification	Less Similar	Developing	
0.30	New Component	New	New to Design Team	
0.40	New Development	New Class for Team	Leading Edge	
0.50	New Dev. & Source	New Class for Industry	Breakthrough	
High values of Alpha normally have an associated high initial failure allocation.				

α Alpha ranges 0.05 to 0.5. Between the line Alpha values are assigned where appropriate. e.g. 0.25 is considered nominal for a well run program. Open ARMS workshop discussions are necessary when setting Alpha. Above normal Engineering actions which increase the rate of learning support an increase in the initial alpha selections. Robust Engineering, FEA, bench tests, Vibration studies, etc.

(SF) Support Factors are based on managerial actions limiting design efforts. e.g.. Overlapping design cycles, evolving specifications, material or supplier changes, decisions to not invest in improving the selected item, test limitations, etc. In some special cases a (SF) of -100% can occur and values of -10% to -30% are not uncommon. Enter comments into input sheet cells.

S F	ENGINEERING SUPPORT RISK	The magnitude of SF is an engineering assessment of support issues
-50%	High	Negative SF's occur when the next Gen of design is committed
-40%		before testing and learning is completed, when new technology
-30%	↑	is introduced or solutions come with there own problems.
-20%		Business conditions may not support all possible improvement actions.
-10%		An SF of 100% means no improvement before the next design cycle,
0%	Low	All solutions can be implemented with little or no new problems introduced

Fig. 5-8

105

ready to allocate risks while in the ARMS' Workshop. All LBS members of the team were present at the start of risk allocations. It is not recommended, but after level ONE and level TWO risk assignments have been agreed on, individuals not connected to or affected by the level THREE sub-group being discussed may step out of the meeting until their areas of responsibility come under discussion. It is not recommended, because the ARMS' process is about sharing knowledge, assessing my assigned risk responsibilities with your assigned risk responsibilities and coming to agreement on where resources will be assigned. If a key engineer is out of the room, it will require a rehash of risk assignments. With some teams, if you leave the room, you can end up with new assignments and no resources :)

It is critical that all members agree and buy-in to the configuration of the ARMS' model as we proceed. We do not want someone to feel left out, only to speak-up later asking for a structural change. If a team member has reservations, they need to be aired as soon as they are known. Some individuals will need help with making their needs known until they become more familiar with the ARMS' process.

The next step is to flesh out the ARMS' structure with estimates of α (Alpha) learning rates and **SF** (Support Factor) issues if they exist **(Fig.5-6,8)**.

The LBS Engineering Team did not know the absolute magnitude of the design risks, and they did not need to know, before assigning items a relative magnitude. What we do know is that all risks in the design add up to 100% and that we can allocate relative risks on a percentage basis from the team's current limited understanding of the design. The chemical engineer played a key role in this assessment. It was important to monitor the c.e.'s requests. Keep in mind, with or without ARMS, engineers will be required to make these same decisions on a daily basis, however, with ARMS the process is visible, may be reviewed and becomes document-ed as a part of "Knowledge Capture **(Fig.5-9)**."

With the ARMS' process, you also know that knowledge capture is in fact taking place and you have an identified location to keep it.

Begin the LBS ARMS' model by pressing the Level_ONE button at the top of the INPUT sheet. In the future, ARMS' models will have a ribbon at the top in place of on sheet buttons. There are VBA, .dll, Com and Scrip based ribbon anomalies and security re-strictions that occur when transferring ribbons between comput-ers that need to be resolved in the ARMS' model before a ribbon approach can be released. Provide me with help if you have any constructive comments or recommendations on which ribbon type to apply to this case and how to proceed.

The LBS Cell design concepts and key manufacturing processes were assigned relative risk percentages across all level ONE ele-ments **(Fig.5-4)**. The process of assigning risk %s must be man-aged by the project engineering leader to assure a sincere effort by all. In this LBS Cell case, the Reliability Engineer made major contributions to that effect. In normal engineering work, the en-gineers are making decisions concerning work assignments and budget allocations from their limited knowledge base with little or no review.

ARMS documents the team's current understanding of the de-sign and aides in developing a risk improvement plan.

It also assists in deciding where specific engineers would be able to make their best contribution. ARMS also aides in identifying engineers with hidden talents who can act as internal consultants on particular issues.

During this LBS Cell ARMS' model it was revealed that a specific technical expertise required was in short supply, because an additional chemical engineer with lithium battery experience was critically needed. The present chemical engineer was working a 60+ hour a week schedule and overloaded with a backlog of requests, while other team members waited for a response.

Obtaining budgetary resources to solve an assigned problem develops a competitive element during ARMS' Workshops and stimulates engineers to discuss and reveal problems and issues that may otherwise lay dormant and emerge later as a surprise and then become a road block in the program. This latency effect is why many programs have an avalanche of late in the development cycle design changes (Fig.Int-1).

Set the ARMS' model to level TWO, then allocate risks across the elements that add up to 100% for each section (Fig.5-5). In this model, a review of knowledge within the team led to assigning 35% of the total level ONE risks to the category FILMS. Keep in mind, you don't know what you don't know. You are allocating your collective understanding at the time. This Lithium Battery Cell project contained a major risk that was little understood during the early stages of the design effort and was only revealed at a later stage. But this fact is OK, the allocations were the team's best judgement at the time and will be corrected as knowledge is gained. There are no wrong answers, only current knowledge.

Throughout the workshop, the team is reminded their assigned risks require budget allocations to be resolved. If an individual engineer thought the risks they were assigned are greater than someone else's, or required a larger share of the budget, this is the time to speak up about the challenge to be solved and what is necessary to solve it. Some problems are more mature and more difficult to improve. In the aviation business, we called mature

LBS ASSIGNABLE ACTIONS

54	DETERMINE REASONS FOR EXCESSIVE GASSING FOLLOWING DEGAS AND FINAL SEAL	Global	Global		Active
56	CORRELATE BICELL IMPEDANCE MEASUREMENTS WITH TYPE 1 AND / OR TYPE 2 FADE.	Global	Global		Active
49	ASSEMBLE ROBUST STUDY ON MIXING AND CASTING OF CATHODE FILM.	85, 86, 87, 88, 89, 95, 96, 97, 178, 25, 44.	24.02		Pending
82	UNDERSTAND CAUSES OF CELL SHORTS - BB PROJECT	119, 129, 140, 188, 189, 190, 191, 192	19.11		Active
38	EVALUATING NEW PACKAGING MATERIALS AND SUPPLIERS TO GET TO 5YR LIFE.	188, 189, 190, 191, 192	18.80		Active
48	ASSEMBLE ROBUST STUDY ON MIXING AND CASTING OF ANODE FILM.	65, 66, 67, 68, 69, 75, 76, 77, 85, 86, 87, 88, 89	17.98		Pending
13	IDENTIFY ATTRIBUTES THAT CONTRIBUTE TO CAPACITY FADE IN CATHODE ACTIVE MATERIALS	85, 86, 87, 88, 89	15.58		Pending
59	DETERMINE CAUSE FOR EXCESSIVE GASSING DURING 65 C BAKE PROCESS	89	10.38		Active
62	UNDERSTAND MIX PROCESS VARIATION DUE TO HUMIDITY	89	10.38		Active
87	EVALUATE HIGHER TEMPERATURE FORMATION AGAINST LOWER WIMP PROFILE	218, 219	10.18		Active
61	INVESTIGATE DIRECT COATING TO COPPER FOIL	25	10.17		Pending
83	EVALUATE PECHINEY ON NEW PACKAGING MATERIAL WITH HIGHER ABRASION RESISTANCE	189	9.94		Active
14	IDENTIFY ATTRIBUTES THAT CONTRIBUTE TO CAPACITY FADE IN ANODE ACTIVE MATERIALS	65, 66, 67, 68, 69	9.44		Pending
78	DEVELOP SEALING PARAMETERS FOR POLYPROPYLENE PACKAGING MATERIALS	169, 172	7.82		Active
12	IDENTIFY POTENTIAL SUPPLIERS FOR COATING GRID/FOIL	14, 15, 16, 17, 18, 19	6.93		Complete
63	EVALUATE SOLVENT SYSTEM COMPATIBLE WITH RESIN ON GRID MIX- ADHESION/RESISTANCE	14, 15, 16, 17, 18	6.93		Pending
81	UNDERSTAND CAUSES FOR STACKED BICELL SHORTS-GB PROJECT	178	5.56		Active
6	SPECIFY THE ATTRIBUTES OF TEMPERATURE RELATIVE TO FORMATION	219	5.09		Complete
43	PERFORM DOE ON ELECTROCHEMICAL IMMOBILIZATION DURING FIRST CHARGE	218	5.09		Active
44	PERFORM DOE ON FORMATION TEMPERATURE VS. PERFORMANCE	219	5.09		
45	DETERMINE OPTIMUM CONDITIONING PARAMETERS FOR PEAK PERFORMANCE	218	5.09		Active
86	UNDERSTAND EFFECT OF FORMATION TEMPERATURE ON PERFORMANCE	219	5.09		Active
88	DEVELOP NEW HIGHER TEMPERATURE NOVRAM OR EQUIVALENT METHOD OF READING AND LOCATING CELL SERIAL NO.	219	5.09		Active
16	DEVELOP AN ELECTRODE THAT YIELD BETTER PERFORMANCE THROUGH THE USE OF HOMOPOLYMER.	197, 198, 199, 200, 202	5.08		Pending
39	PERFORM ELECTROLYTE VS. PERFORMANCE STUDY FOR ACCEPTABLE PERMEATION.	197, 198, 199, 200, 202	5.08		Pending
41	PERFORM DOE ON ELECTROLYTE ADDITIVES VS. PERFORMANCE	197, 198, 199, 200, 202	5.08		Complete
52	SPECIFY THE OPTIMUM ACTIVATION ATTRIBUTES (VACUUM, ELYT. QUANTITY, ETC.)	197, 198, 199, 200, 202	5.08		Complete
33	DETERMINE USEFUL LIFE OF CAT DESIGN, WRT LONG TERM PERFORMANCE	167, 169	4.89		Complete
75	DEVELOP ENVIRONMENTALLY FRIENDLY SURFACE PREPARATION THAT'S STILL EFFECTIVE IN SEALING CELL	169	4.76		Active
79	EVALUATE ALTERNATIVE PLATING TECHNIQUES ON TERMINAL SURFACE PREP.	169	4.76		Active
57	UNDERSTAND IMPACT OF LAMINATION PROCESS ON FILM CONDUCTIVITY AND PERFORMANCE	117, 118, 119, 127, 128, 129	3.65		Complete
58	UNDERSTAND DIMENSIONAL VARIATION DURING BICELL LAMINATION (INCLUDES MQ1 AT SUPPLIER)	117, 118, 119, 127, 128, 129	3.65		Complete
91	UNDERSTAND SEAL STRENGTH DETERIORATION VS LIFE OF CELL	239, 191	3.53		Active
77	DEVELOP TERMINAL TAPES APPROPRIATE FOR POLYPROPYLENE PACKAGING MATERIALS	172	3.05		Active
15	IDENTIFY ATTRIBUTES THAT CONTRIBUTE TO CAPACITY FADE IN OTHER NON-ACTIVE MATERIALS	78, 98	2.54		Pending
8	ADD ADHESION TO M-SPEC FOR COATED GRID/DEVELOP TEST PROCEDURE FOR ADHESION	37	2.38		
60	INVESTIGATE DIRECT COATING TO ALUMINUM FOIL	44	2.37		Pending
29	PERFORM STUDY ON SEALING OF THICKER TERMINALS	191	2.31		Complete
40	PERFORM DOE ON CELL TO DETERMINE PEEL STRENGTH WITH ELECTROLYTE	191	2.31		Complete
55	UNDERSTAND ELECTROLYTE IMPURITIES TYPES AND LEVELS AND THEIR IMPACT ON PERFORMANCE	202	2.14		Pending
74	UNDERSTAND EFFECT OF PERFORMANCE VS. ELECTROLYTE IMPURITY LEVELS	202	2.14		Pending
85	EVALUATE ELECTROLYTE PURITY AND LEVEL VS. PERFORMANCE	202	2.14		Active
4	SPECIFY THE ATTRIBUTES OF A SUITABLE CATHODE FILM MIXTURE & THE MEASUREMENT SYSTEM FOR PROVING IT	100	2.08		Pending
51	EVALUATE THE CONTRIBUTORS TO HOMOGENEITY OF CATHODE FILM.	100	2.08		Pending
3	SPECIFY THE ATTRIBUTES OF A SUITABLE ANODE FILM MIXTURE & THE MEASUREMENT SYSTEM REQUIRED.	80	1.46		Pending
50	EVALUATE THE CONTRIBUTORS TO HOMOGENEITY OF ANODE FILM.	80	1.46		Pending
5	SPECIFY ATTRIBUTES (CLEANLINESS) OF A SUITABLE GRID MATERIAL & MEASUREMENT SYSTEM REQUIRED, ADD TO M-SPEC.	26	1.34		Complete
47	INVESTIGATE CLEANLINESS OF BARE GRID MATERIALS IN TERMS OF ITS IMPACT ON COATING AND WELDING.	26	1.34		Active
90	UNDERSTAND EFFECT OF SELF DISCHARGE ON CALENDAR LIFE	240	1.22		Pending
36	PERFORM TOLERANCE STACK-UP OF BICELL IN STACK & WELD FOR PACKAGING,	171	1.17		Complete
37	PERFORM TOLERANCE STACK-UP OF BICELL IN STACK & WELD FOR SYSTEM APPLICATION	171	1.17		Complete
73	DEVELOP METHOD TO PREVENT TERMINAL EDGES FROM CAUSING CANCER	171	1.17		Active
76	DEVELOP PM SCHEDULE TO PREVENT BURRS ON TERMINALS	171	1.17		Active
80	VARY BURRS AND UNDERSTAND EFFECT ON FAILURES FOR DIM SPECS	171	1.17		Active
65	EVALUATE XXXXX AS SEPARATOR SOURCE	136, 137, 138, 139, 140, 141	1.01		Active

Fig. 5-9

problems "Hanger Queens," the problems that never made it out to fly. Under such conditions, conducting a design of experiments study, contracting outside services, hiring additional specialized personnel, and the purchase of specialized capital equipment may be required. The risk of not solving a mature problem may not be high, but mature problems will adversely affect our α (Alpha) learning rates and **SF** (Support Factors). There may be a simple low risk issue that requires a major capital investment to solve. The risk level assigned is still low, but the Support Factor may be high, requiring the budget issue to be reconciled outside of the ARMS' model. The LBS Cell development project required several D.O.E. sidebar efforts. The LBS management even brought in Mr. Shin Taguchi to train and coach the LBS engineers in D.O.E. techniques. Management amortized the cost across anticipated future business. The ARMS' model often brings conditions of mature problems not being solved to the surface and gives management the option to resolve a solution path. LBS Cell development was new enough that mature problems had not yet developed. There were some ignored problems due to the resource limitations and priorities, however.

This brings to mind a case from the early days of commercial jet aircraft. The Boeing 727 was about to be released and FAA trials were being completed in Miami, Fla. I was working on an MBA and was asked by Mr. Jim Barfield, founder and owner of Barfield Instruments, to conduct a business case study on whether or not he should invest in obtaining digital test equipment to service new digital instruments the 727 planes would have. Mr. Barfield's present business included servicing analog instruments for several regional airlines. In conducting the financial study, I interviewed the CFO, Marketing Manager and Chief Engineer to obtain their understanding of the new opportunity. After a full review and documentation of the financial numbers, it was clearly not a feasible direction for the company. I knew Mr. Barfield would be disappointed with my report. He was an entrepreneur with "fire in his belly," who had built his business from the ground up. I sat with Mr. Barfield and his CFO for the final report discussion. Before handing it over, I asked Mr. Barfield why he thought he should consider the new approach. His eyes lit up and he talked about the future of aviation and the unidentified business to come. Listening to him was inspiring! I said "Mr. Barfield, my 'analytical financial study' clearly says you should not buy the equipment, but numbers are numbers and your insight transcends that. You should follow your passion and go for it." Mr. Barfield did just that, and ordered the over one million dollars worth of digital test equipment. He was absolutely right! When the equipment was delivered, he arranged to pay the truck driver with a check, before the driver was allowed to unload the equipment. The notoriety of his actions spread quickly. None of the commercial airlines had purchased similar equipment. In fact, when they heard what Mr. Barfield had done, they delayed or cancelled their test equipment procurements then signed service contracts with Barfield Instruments. With that single stroke, Mr. Barfield cornered the digital aircraft instrument repair market for several years and made a fortune. The point of telling this true story is that engineers should not sell management's role short. An entrepreneurial manager can come up with solution paths that an engineer never would or ideas that would be rejected by an engineer because, quite frankly, the decisions are above the engineer's pay grade.

An ARMS' Workshop is an opportunity for the whole team to learn about the broader challenges the project and company are facing.

Team members not connected to a risk frequently can suggest solutions, based on their prior experience, and help someone else mitigate an assigned risk without needlessly drawing on limited budget resources.

Frequently check to make sure the ARMS' Document Engineer assigned to capture risk issues is doing so, and is letting the team see what is being captured as each item is discussed.

SAVE "Ctrl+S" the model frequently. From time to time, save with a new name such as an added -#. This will leave a historical copy at that point in time in case it is needed for comparison or reset later.

Many issues can arise during a meeting that are not design items. These should be noted on the side or passed on and addressed later as deemed appropriate. These issues may include problems with purchase orders or other business systems, for example. I have even had a broken coffee machine come up as an item affecting team performance. Note that while allocations of risk are being assigned, and the total of percentages will be indicated on the INPUT page, they must add to 100%. In addition to the risk percentages already assigned, at level THREE, the team will estimate α (Alpha) learning rates and **SF** (Support Factors). As each level THREE **(Fig.5-6)** risk is discussed and assigned, it is usually best to also assign them α and **SF**. This will avoid rehashing prior rationale and shorten the total time to make assignments.

An ARMS' guide for Alpha and Support Factors is provided **(Fig.2-16 & 5-8)**. Each team member should have their own reference copy to review during the workshop as they make assignments. In this LBS Cell ARMS' Workshop, Acrylic Adhesive received an α of 0.35, an indication that they were early in the learning process for this line item. Acrylic Adhesive also received a Support Factor of − 20%. This means that 20% of the possible learning is forecast to be unrealized or lost. There can be many reasons for this; limited sample builds, late sample results, technician shortages, problems with analysis equipment and so on. I strongly encourage that the reasons for all **SFs** be entered into the ARMS' model as cell comments. Remember, **SFs** can only be negative and are resolved through management actions. During reviews, they will and should be questioned. **This is where the team has an opportunity and obligation to document anything which is interfering with their ability to be successful in their work.** A lack of equipment, personnel, time, supplier development, supplier support and excessively long procurement cycles are typical impediments to be clearly documented here.

Amazing things have been brought up under **SF** number assignments. The LBS Cell project was no exception. If there are any dysfunctional business systems, the engineering community is aware of it and will discuss the impact. For example, these might include purchase order approvals bogged down by bureaucratic hurdles and even engineers not getting support because they never asked for it. I refer to the **SF** column as "Management's Column" and request management to fully understand anything placed in it. During the management review, after the ARMS' model is built, there needs to be reconciliation of **SFs** with the engineering team and **SF** improvement plans made. This action always makes management a closer part of the team. A **SF** should not be lowered by increasing a risk %. Management owns the **SF** and management needs to bring **SFs** to a resolution. If there is a new design approach which eliminates the need for an **SF**, that is a different matter.

After Risks, α Alpha and **SF's** are made for (GEN 1.) Risk values for (GEN 2) are automatically calculated based on (GEN1) α Alpha and **SF's** when the "UPDATE" button on the INPUT sheet is pressed.

In all cases, determining the effect of changes made, requires pushing the "UPDATE" button on the INPUT tab **(Fig. 2-7)**. That is a good time to also press **CTRL+S.**

Alpha values typically diminish from one generation to the next as quick and easy actions are being resolved and the anomaly is maturing. **SFs** are, hopefully, improving. This may not be true in all cases and the reasons for unusual Alpha and SF values need to be entered into cell comments. For example, the need for a new piece of equipment may be recognized. These comments become invaluable during management reviews or your own review in the following months. Remember the "Dr. Hermann Ebbinghaus Effect." During the later stages of the LBS Cell development, delayed resources due to the unannounced sale of the division began to surface. Concurrent with Level_THREE α Alpha and **SF** assignments, discussions of actions are captured as **'Assignable Actions' (Fig. 5-4)** by the ARMS' Documentation Engineer. The lead engineer should assign them to an individual for solution at the same time and seek an estimate for completion. This is when you find out if individuals are truly overloaded. The LBS chemist was a classic example of the effect on team performance due to individual overloading.

Do not be overwhelmed by all of the data being captured. With a little management, the ARMS' Workshop moves quickly. Usually the team is well trained by this point in the ARMS' Workshop and they can take over and pick up the pace of the meeting. The lead engineer needs to make sure the integrity of the process is maintained. With the LBS Cell project, it was the engineers who would call for a new meeting about every 3 months because they recognized they were beginning to drift in their efforts or that **SFs** were becoming an issue.

It is critical to the ARMS' model that "comments" entered on

114

proposed solutions be discussed during the ARMS' Workshop while the team is present. Do not delegate the generation of Assignable Actions to an off-line responsibility of the ARMS' Documenting Engineer or any individual. The discussions reveal more actions than a single person would be able to develop working "off line" and the actions need to be reviewed by the team as they are generated. I have found that teams accustomed to **FMECAs** tend to make this error of "off line" delegation. The object is not to offer up a completed ARMS' model. It is to have the interaction between team members. Interaction leads to discovery and commitment to solution paths that are made while the ARMS' model is being built. A completed ARMS' model is a piece of paper or screen shot. The commitment and active improvement plan is what is important.

In an ARMS' model, the Assignable Actions are listed in any order as they come up during discussions. They do not need to be lined up across the INPUT sheet. Documented actions are coded with the several "item line numbers," which will benefit from the particular action in question. Several actions may be assigned to the same "item line number" and vice versa. The line numbers are entered with a comma space delimiting pattern. IPT (Incidents per thousand) and the SF indexes are automatically calculated from the sum of their contributors.

Diligently making the line # assignments is very critical to the success of the ARMS' Weibull forecast. This is a difficult step which requires discipline from the team, so do not delegate this task.

Line # assignments include personnel assigned to an issue, status of activity and any additional comments to clarify what needs to be accomplished and they will be updated between workshops as the program proceeds. Later, when ARMS' reports are generated, the Assignable Actions will be sorted by risk reduction potential and placed into a Pareto listing.

Three important reports are generated by ARMS;

(1) **Assignable Actions & Status,** by descending risk potential.

(2) **Design Elements,** by descending risk potential.

(3) **Weibull Analysis** forecasts, based on Alphas, SFs and Assigned
Actions with Validation Test results.

Failure Rate, Reliability Growth and MTTF Growth graphs are also
developed. All are developed from the same data, and with experience can be determined directly from the Weibull representation.

These reports indicate which assignable actions have the largest
return on investment (Bang for the Buck), and which items are
the highest risk, the Achilles Heels. ARMS' reports are reviewed at
the closing phase of the ARMS' Workshop to make sure there is
team agreement, that the reports make sense and reflect what
they believe to be accurate. If a team member has a problem with
any part of the ARMS' reports, their objection needs to be discussed and resolved. These reports are a strong input to the creation of a "Risk Improvement Plan."

ARMS' reports and ARMS' models are knowledge capture!

When and if there are future design iterations, the ARMS' model
is valuable information to review. ARMS' information also develops a degree of stability for the project as new information becomes available with development progress. The Assignable Actions are also a valuable resource when planning Design of Experiments studies. When using the ARMS' reports, do not run down
the Pareto of Assignable Actions top to bottom **(Fig.5-9)**. The listing is a guide to assist the Project Leader in making the best use

of time and resources, it is not a punch list or step by step plan and they should not be used in that manner. Between ARMS' Workshops, many new problems will be revealed and need to be considered in light of the ARMS' reports. ARMS' reports are living documents supporting the project leadership and require active scrutiny.

From the LBS Cell ARMS' model, input data was entered into a Weibull plot (Fig.5-11). Lines for original conditions, (GEN 1), (GEN 2) and (GEN 3) were made based on risk improvements projected by the assigned α Alpha and **SF** Support Factors.

The ARMS' software has compiled all data from discrete projections to show an expected risk improvement. Keep in mind the projection will only become viable if the work assigned is actually performed and completed. It is normal for some of the assigned work to still be in progress when the next ARMS' Workshop is scheduled. During the next ARMS' Workshop progress on Assigned Actions can be updated and clarified.

As the LBS Cell development continued through (GEN 1) to (GEN3); a new Build and Validation Test program from each generation was conducted. At each major design iteration, a new ARMS' model was constructed. Validation Test sample sizes were 9, 18, and 36 individual lithium battery cells. Engineering development continued throughout the test period and many improvements were made in between formal design generations that are not discussed here.

Knowledge gained through the analysis of validation test cells and the arrival of new capital equipment led to additional process improvements. Validation Tests were in an overstress condition based on higher than normal charging currents and more rapid discharge rates than would be encountered in normal usage.

Higher internal cell currents resulted in internal I^2R heating. This required close monitoring to keep heat below an upper threshold set by the Lithium Battery chemist. Additional external environmental control was required. A Validation Test **overstress of 1.2 and 1.2 lifetimes** was set based on more charge discharge cycles/day than during normal life and a longer time on the test stand. The build %σ was believed to be about 33%.

6/9 GEN 1 (**Fig. 5-10 & 11**) Nine LBS Cells tested had failures at 50, 290, 486, 620, 702, and 840 cycles of charge-discharge with three tests being suspended. The Weibull GEN 1 line was calibrated using the new data.

The "INITIAL" design **START** point failure rate was reset to 370 IPT (37%) failures per year. Placing the "knee" of the bath tub on the GEN 1 Forecast line at 210 IPT (**Fig.5-11**). Failure analysis yielded several abnormal conditions requiring Assigned Actions during GEN 2. This was a greater rate of failure than the team expected and was disappointing.

Manufacturing processes, which occurred under Engineering Lab

(Test 1)			γ Gamma Correction =		Program Name			
TOTAL	9		1	< Acceleration Factor if any			Beta Test_1 >>	1.80
Line #	QTY TAKEN OFF TEST	Failure=1 Suspension=0	Rank Individual TIMES or CYCLES at Test Termination	INCREMENT	Mean Order Number	Previous Mean Order Number	Median Rank %	
			CYCLES		1.00			
1	1	1	50	1.0000	1.00	8.00	7.413%	
2	1	1	290	1.0000	2.00	7.00	18.059%	
3	1	1	486	1.0000	3.00	6.00	28.706%	
4	1	1	620	1.0000	4.00	5.00	39.353%	
5	1	1	702	1.0000	5.00	4.00	50.000%	
6	1	1	840	1.0000	6.00	3.00	60.647%	
7	3							
8								

Fig. 5-10

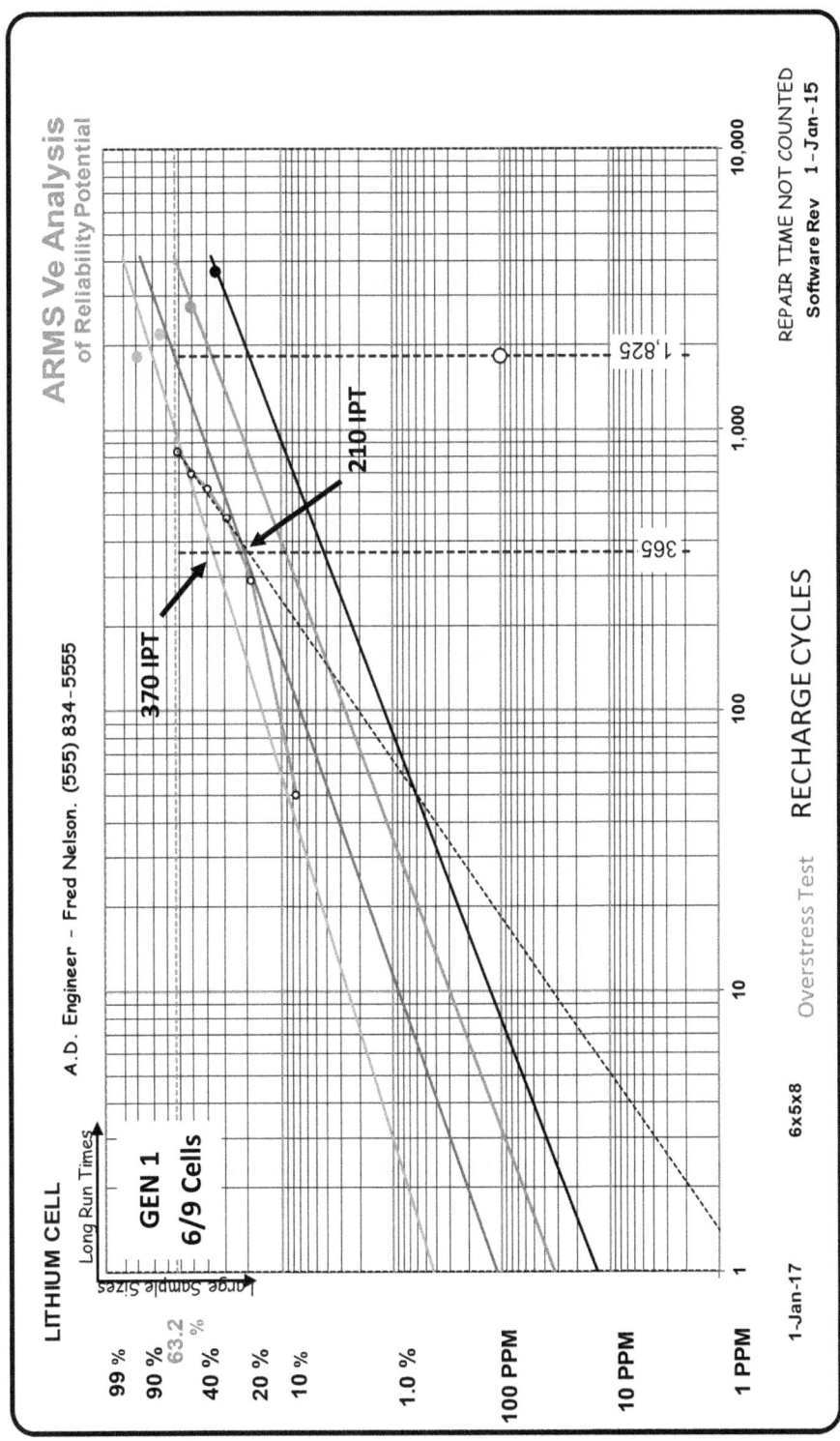

Fig. 5-11

119

conditions, were a major contributor to the lack luster performance. As a result of the analysis findings and test results, the team that the new forecast was a good estimate of where they were in the design effort.

Initially, the team thought they would all fail in the first year. The initial **START** value had been set to an ARMS' default value of 500 IPT. The test result of 370 IPT was actually better than the LBS Engineering Team felt.

This high initial calibrated failure rate of 370 IPT contains failures that need to be classed as premature. The premature nature is reflected in an "Initial" Weibull line with a beta slope of less than one followed by three generations of design with beta slopes approaching one. The resulting "Initial" Weibull line is divergent with the GEN 1-3 projections. The "tipping" of the β line can be observed on the plot at the high right end when compared to the DOTs **(Fig.5-11)**.

Failure analysis did clearly reveal one premature failure mode due to moisture and a sealing issue. There also were five early failures in an apparent end of life condition. This is indicated by the upward slope of the Weibull graph plot. As a result of these findings, renewed effort was put into the cell sealing process and material moisture control and handling issues on the prototype production line. All processes pertaining to material purity were also reviewed and a few adjustments made. Several configuration improvements were also made based on the failure analysis and fresh ideas from the engineering team. The **START** value for (GEN 2) was adjusted to reset the "INITIAL" line to 210/1000 **(Fig/5-13)**.

6/18 GEN 2 **(Fig. 5-12 & 13)** Eighteen Cells from the GEN 2 design and processes were submitted to durability testing. The testing was stopped after six failures before all units were destroyed, permitting analysis of cells that had been stressed but had not failed. This was important to understanding the nature of more successful cells and to learn from cells that had not been destroyed in the process of failing. Impending failure evidence led to a few construction design/process details being improved.

Adjustments were made to the production processes based on failure analysis results and on recognizing the importance of improving control of production line conditions. These results and specific actions will not be discussed in detail.

Also notice what is happening when applying a Weibull approach.

The Engineering Team's thoughts and expectations are being plotted on the same graph as measured results!

This provides valuable feedback to the team and helps them reorient their understanding of the state of their development

One condition that has since become common knowledge in the lithium battery industry will be discussed.

It was suspected, during an ARMS' Workshop, that the lack of production line moisture control was playing a major part in the lithium battery failures. Surface moisture and moisture entrained in hydroscopic battery materials can be released by charge/discharge related I^2R heating. Trace moisture freed from barrier films can lead to lithium chemical interactions within the battery causing a runaway chain of events; with

6/18 GEN 2			$\Gamma 2 =$		Program Name		
TOTAL	18				Beta Test 2 >>		4.5
Line #	QTY TAKEN OFF TEST	Failure=1 Suspension=0	Rank Individual TIMES or CYCLES at Test Termination	INCREMENT	Mean Order Number	Previous Mean Order Number	Median Rank %
2			CYCLES		1.00		
1	1	1	205	1.0000	1.00	17.00	3.778%
2	1	1	1,005	1.0000	2.00	16.00	9.216%
3	1	1	1,135	1.0000	3.00	15.00	14.653%
4	1	1	1,275	1.0000	4.00	14.00	20.091%
5	2	1	1,392	1.0000	6.00	12.00	30.967%
6	12		1,675				
7							

Fig. 5-12

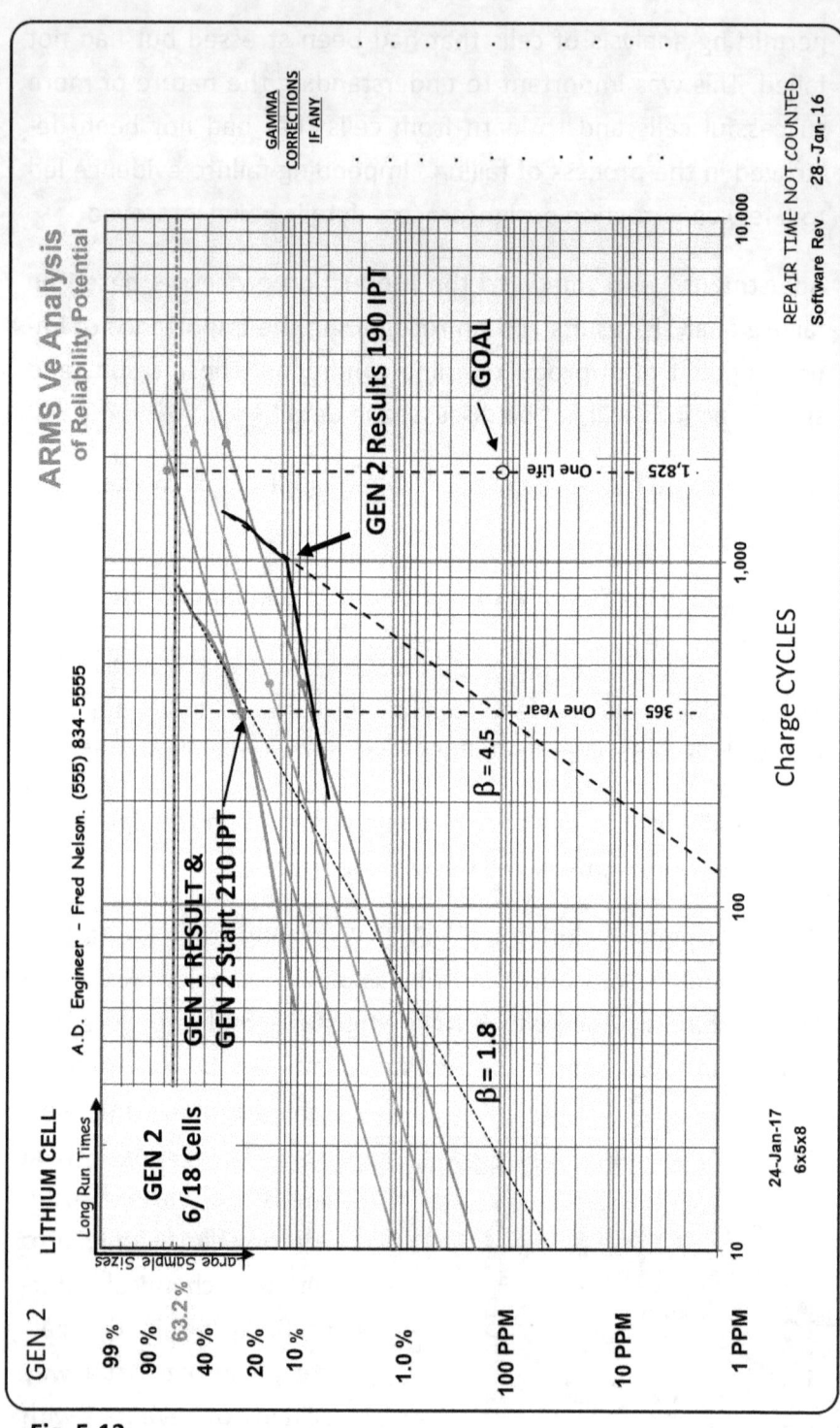

Fig. 5-13

additional heat generation and additional moisture release from all sources within the cell.

Batteries were being built in an engineering lab on equipment that was being changed out for production grade equipment as the new equipment became available. Production processes were being developed by the manufacturing team concurrent with LBS Cell design developments. These diverse engineering actions were brought together by mutual participation in ARMS' Workshops.

The assembly area was air conditioned, however, there was a significant variation in humidly levels as the week progressed due to people activity through the week and weekends. A series of improvements were taken as the durability test builds continued. At this point in the development, a customer was also conducting their tests on our prototypes. Some of the corrective Assignable Actions from the ARMS' Workshops were:

1) Assembly room doors and windows were sealed and access controlled.

2) Humidity measurements were logged and humidity levels controlled.

3) Autoclaves were employed to dry all materials at supplier locations before shipment and again immediately prior to assembly.

4) Sealed moisture resistant storage containers with desiccators for supplier shipments and inventory storage with first in first out controls.

5) Storage of all inventory materials in a humidity controlled environment prior to staging in the assembly room pre-

production.

6) Increased bake times and temperatures at lower oven humidity levels. Improved oven exhaust systems.

7) Q.C. sampling for material moisture content before, during and after battery assembly using new special humidity test equipment.

8) Linking between production line moisture data, time stamps and lithium battery serial numbers.

9) Link assembler work schedules and assignments to battery serial numbers. No skin contact with materials during handling or assembly and elimination of hand creams and moisturizers.

10) Review all production line moisture data during analysis of durability test failures.

This discussion only covers the highlights of activity that occurred between ARMS' Workshops as the design progressed.

ARMS' Workshops renewed focus on critical activities while informing the full team, including management, on progress being made. In some cases, sidebar studies of sub-components and processes were conducted between design generations and the knowledge gained was incorporated when feasible.

With ARMS, you have the ability to add a vector simulating a beta slope starting at the highest test data point. This is achieved by entering an angle value into the "Validation tab" box, next to the Beta Test_#. The beta angle value entered is adjusted by trial and error until the slope of the vector line matches the slope of the wear out portion of the bath tub. In this case a slope of 4.50 fits the data in wearout mode **(Fig.5-13)**. The vector projects the wearout portion of the bath tub curve down to indicate an estimate of where wearout would be encountered with a larger test group.

This is an estimate of the median values which have a wide range interval due to the small sample size. The bottom of a bath tub, which would be revealed with a larger test group, would be lower on the graph and intersect with the beta vector, our estimated characteristic wearout line. Test GEN 1 units are not forecasted to survive to the right of the wear out vector line. Through redesign, we need a characteristic wearout vector that intersects or is to the right of our goal, 100 PPM at 1825 cycles **(Fig.5-13)**.

More than three months passed between the initial ARMS' Workshop and the second workshop and three additional months before a third workshop. As stated earlier in this book, in the automotive world it is common for a new design concept to take in the order of 12 budgeted design cycles prior to production. ARMS improves the effectiveness of each design cycle, resulting in lower total cost and taking less time. The cost improvements are primarily from getting the right answers sooner and terminating waste.

Knowledge gained in each LBS design generation was used to make a new ARMS' Workshop model. In this example, for brevity, we are only showing the Assignable Actions from the latest ARMS' model.

At each new design generation, a new ARMS' model needs to be constructed. Do not try to flow GEN 1 into GEN 2 and GEN3 in the same model. You would waste time shuffling the deck chairs on the Titanic. It is faster and more effective to build a new model. I like to think of the future Weibull projections as pinging the future like sonar. Your new models can be labeled GEN 2, GEN 3 and

GEN 4 or as you like. It is important to keep the ARMS' reports on Design Items and Assignable Actions refreshed. You can abridge uncompleted actions from earlier models into the new model if you update the line #'s for the new model. The reports document progress and bring clarity to the challenge remaining. Reports from earlier ARMS' models should be kept for future reference. The prior model is used as a guide and can speed up new workshop discussions. It can also be useful as captured knowledge during a case review. Usually there are discoveries during the three months that were never anticipated.

After the third ARMS' Workshop improvements were designed into the lithium cells and additional capital equipment, purchased for production, were put into the engineering prototype manufacturing process. Test units became less costly and larger test quantities became available. Improvements in test stands and a larger quantity of test stations were brought on line. With the Trumpet chart in mind (Fig.1-6), we did not need a large sample of test units, however, there was concern that random nonlinear variations in manufacture could distort results. A larger test quantity would help in identifying if variations of this type were present. Because the design was actively being changed, a very large sample would be

15 / 36 GEN 3			γ Gamma Correction =			Program Name		
TOTAL	36						Beta Test_3 >>	4.50
Line #	QTY TAKEN OFF TEST	Failure=1 Suspension=0	Rank Individual Measures at Test Termination	INCREMENT	Mean Order Number	Previous Mean Order Number	Median Rank %	
			CYCLES		1.00			
1	1	1	1,132	1.0000	1.00	35	1.907%	
2	1	1	1,382	1.0000	2.00	34	4.655%	
3	1	1	1,566	1.0000	3.00	33	7.403%	
4	2	1	1,674	1.0000	5.00	31	12.900%	
5	3	1	1,801	1.0000	8.00	28	21.144%	
6	3	1	2,106	1.0000	11.00	25	29.389%	
7	4	1	2,306	1.0000	15.00	21	40.381%	
8	21							

Fig. 5-14

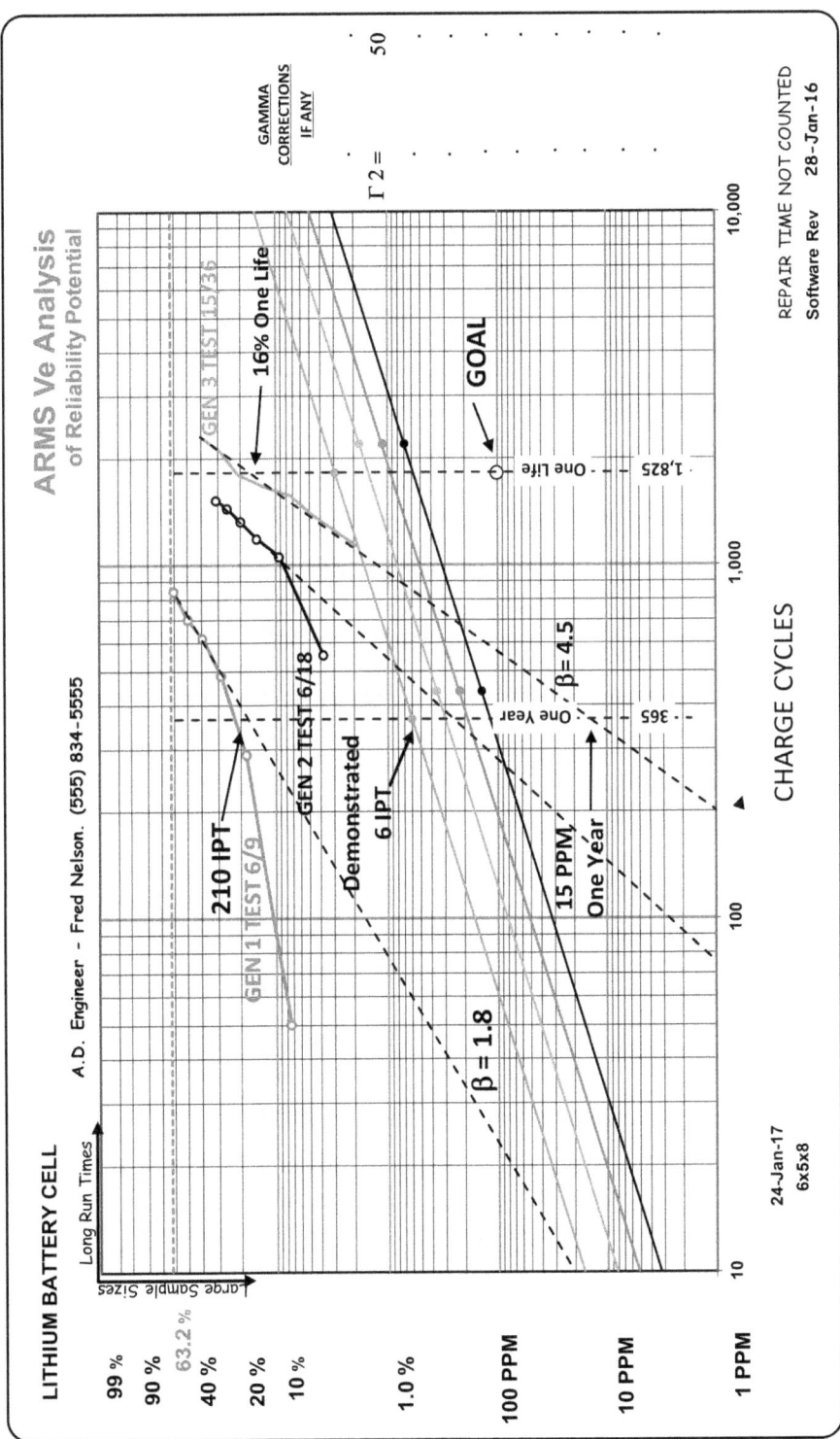

Fig. 5-15

wasteful at this point in the development.

15/36 GEN 3 (**Fig. 5-14 & 15**). Thirty-six cells from a preproduction run of several hundred were selected and tested. As seen in the Weibull graph **(Fig.5-15)**, the wearout line was shifted to the right more than forecast. With analysis of cells that had failed and had survived testing, the improvement benefited from the inclusion of new production equipment to make the pre-production units and critical design improvements in chemistry and component geometry. While these improvements were impressive, clearly several additional design cycles or breakthroughs were needed.

Another common use of the Gamma Adjustments occurs when units are durability tested and the survivors are taken off the durability stands for performance testing, then returned to the durability stands. The original durability period and any additional operation during performance test become a value that needs to be entered as a Gamma function. In the case of GEN 2 **(FIG.5-15)**, a Gamma adjustment of 50 hours was required to account for operation prior to the start of the validation test.

(Fig. 5-16, 17, 18 & 19) Validation test data often contain an early failure near the start of the test and midway through the test during a period considered to be part of normal life. These failures are of equal importance, but for the purposes of isolating common cause sets of failures, the data can be decomposed. With ARMS, we decompose visually using GUI (Graphical User Interface) techniques. The data in **(Fig.5-16,18)** have had individual data points removed, after reviewing the Weibull graph of **(Fig. 5-15)**. In GEN 1 and GEN 2 data, the first failure in each was removed from the set. They were analyzed and found to be due to proto-type internal electrical connections. After removing the two units, the failures each form a straight line as shown on **(Fig. 5-19)**. Also note that both GEN 2 and GEN 3 results now appear to have a β slope

<table>
<tr><td colspan="3">6 / 9 -1   GEN 1</td><td colspan="2">Γ 1 =</td><td colspan="5">Program Name</td></tr>
</table>

Line #	QTY TAKEN OFF TEST	Failure=1 Suspension=0	Rank Individual TIMES or CYCLES at Test Termination	INCREMENT	Mean Order Number	Previous Mean Order Number	Median Rank %	Plot Point Y	Plot Point X
TOTAL 8			1	< Acceleration Factor		Beta Test_1 >>	2.2	1	840
1			CYCLES		1.00				
1	1	1	290	1.0000	1.00	7.00	8.300%	-2.4460	290
2	1	1	486	1.0000	2.00	6.00	20.214%	-1.4880	486
3	1	1	620	1.0000	3.00	5.00	32.128%	-0.9479	620
4	1	1	702	1.0000	4.00	4.00	44.043%	-0.5437	702
5	1	1	840	1.0000	5.00	3.00	55.957%	-0.1984	840
6	3							-0.1984	840
7									
8									

Fig. 5-16

<table>
<tr><td colspan="3">6 / 18-1 GEN 2</td><td colspan="2">Γ 2 =</td><td colspan="5">Program Name</td></tr>
</table>

Line #	QTY TAKEN OFF TEST	Failure=1 Suspension=0	Rank Individual TIMES or CYCLES at Test Termination	INCREMENT	Mean Order Number	Previous Mean Order Number	Median Rank %	Plot Point Y	Plot Point X
TOTAL 17			1	< Acceleration Factor		Beta Test_2 >>	4.5	1	1,475
2			CYCLES		1.00				
1	1	1	1,005	1.0000	1.00	16.00	3.995%	-3.1997	1,005
2	1	1	1,136	1.0000	2.00	15.00	9.746%	-2.2775	1,136
3	1	1	1,275	1.0000	3.00	14.00	15.496%	-1.7815	1,275
4	1	1	1,392	1.0000	4.00	13.00	21.247%	-1.4319	1,392
5	1	1	1,475	1.0000	5.00	12.00	26.998%	-1.1562	1,475
6	12							-1.1562	1,475
7									
8									

Fig. 5-17

<table>
<tr><td colspan="3">15 / 36 GEN 3</td><td colspan="2">Γ 3 =</td><td colspan="5">Program Name</td></tr>
</table>

Line #	QTY TAKEN OFF TEST	Failure=1 Suspension=0	Rank Individual Measures at Test Termination	INCREMENT	Mean Order Number	Previous Mean Order Number	Median Rank %	Plot Point Y	Plot Point X
TOTAL 36			1	< Acceleration Factor		Beta Test_3 >>	4.5	1	2,306
3			CYCLES		1.00				
1	1	1	1,132	1.0000	1.00	35.00	1.907%	-3.9500	1,132
2	1	1	1,382	1.0000	2.00	34.00	4.655%	-3.0435	1,382
3	1	1	1,566	1.0000	3.00	33.00	7.403%	-2.5650	1,566
4	2	1	1,674	1.0000	5.00	31.00	12.900%	-1.9797	1,674
5	3	1	1,801	1.0000	8.00	28.00	21.144%	-1.4374	1,801
6	3	1	2,106	1.0000	11.00	25.00	29.389%	-1.0556	2,106
7	4	1	2,306	1.0000	15.00	21.00	40.381%	-0.6593	2,306
8	21							-0.6593	2,306

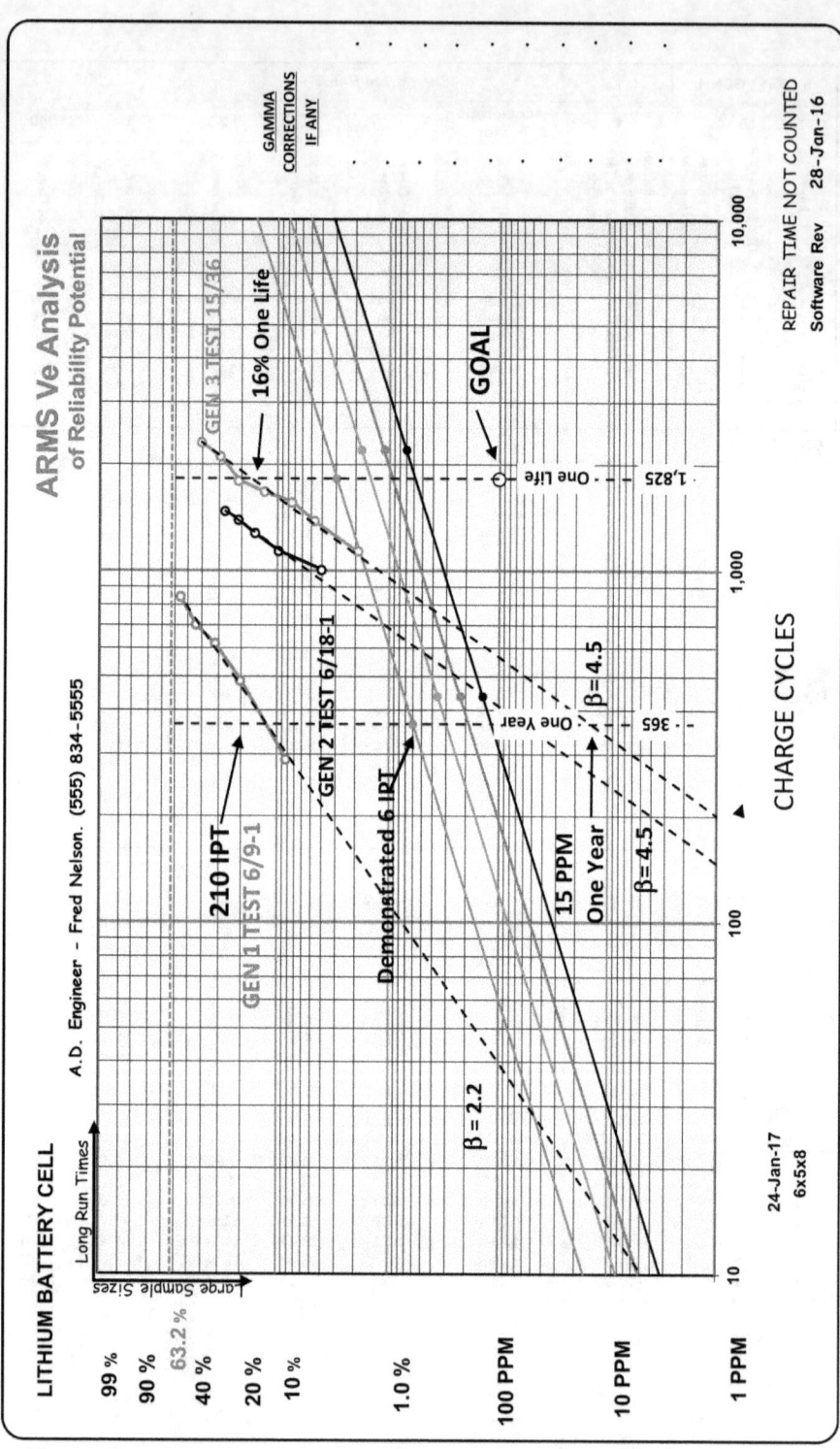

Fig. 5-19

130

of 4.5. The data is suggesting a common cause that improved be-tween GEN 2 and GEN 3. There can be other explanations, how-ever, and thorough investigation of failure conditions is required

During an evaluation of 10 non-thermal plasma reactor units sur-vived 150 hours of durability testing. Then they were taken off of the durability stands for performance testing. All units passed requirements and they were returned to the durability stands for testing until failure.

This is a time when a Gamma adjustment needs to be made. (Reliability Engineering Handbook, Dimitri Kececioglu ISBN 0-13-772294-X). A Gamma adjustment takes into account expended life prior to the start of life testing. Caution needs to be exercised when using the Gamma function to not make excessive adjust-ments which can unreasonably distort your results. A Gamma adjustment must pass commonsense judgement and be con-sistent with the spirit of Dr. Fermi solutions. The cause for adjust-ment should be identified and documented in your model.

The test data from the ten units was entered into an ARMS' mod-el on the Validation Time Tab. As a result; the failure plot was distorted by the unreported original test time (Fig.5-20). The β slope of the result was estimated at 2.5 by trial and error by mak-ing entrees in the Validation Time data sheet.

In case the Gamma value was set to 150 Hours based on the pri-or durability test time expended. Several things occur as a result of the adjustment, the measured data curve shifts to the right by the number of hours we added, the tail of the plot straightens as an affect of the Weibull log scales and is now on the slope of β 2.5. On the right edge of the Weibull graph, and a Γ 2 symbol ap-pears with a value of 150 hours. Anytime a Gamma value is ap-plied, the Gamma value automatically is identified and displayed on the right edge of the Weibull graph. This assures that values

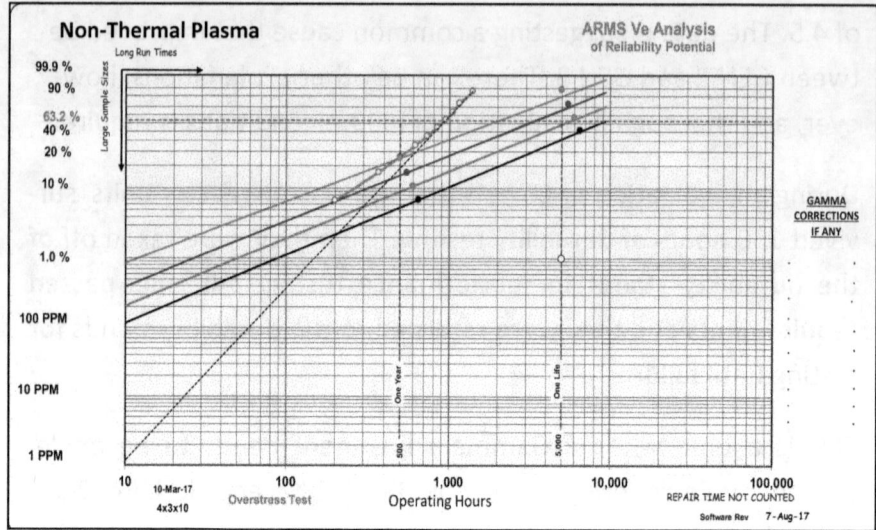

Fig.5-20

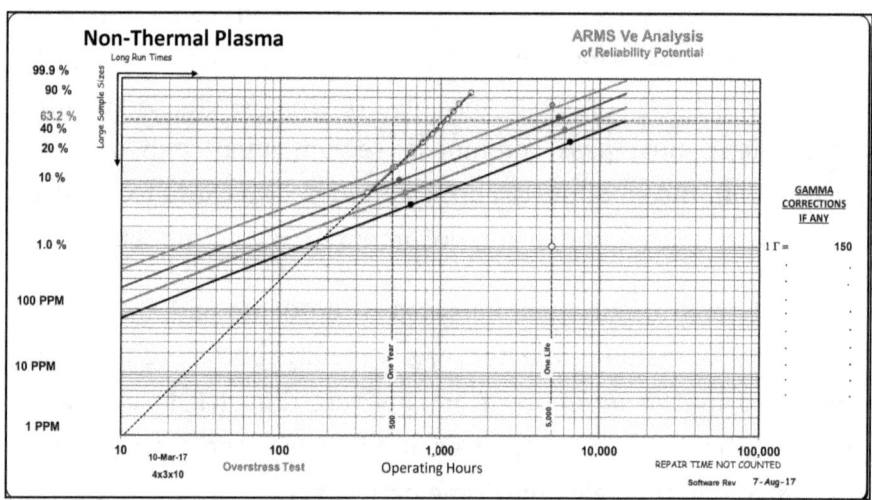

Fig.5-21

unintentionally entered are not hidden:)

Keep in mind, the biggest contributions from ARMS are: **Open Team Interactions, Knowledge Capture, a Pareto of Risk Items, and a Pareto of Assignable Actions which are inputs to forming a Project Plan and Exit Gate strategy.**

Larger development projects benefit from a building block approach, wherein suppliers, or sub-development teams, have a separate ARMS' Workshop on their section of the program. Selected program engineers should participate with key suppliers at the supplier's location. This close interaction brings a multitude of benefits to the program by significantly improving communications at a detailed actionable level.

Supplier ARMS' Workshops provide important casual communications and the reading of body language. It also gives you a chance to pick up the bar bill while bonding your team relationship. In large Fortune 500 companies, this is usually a problem due to internal IRS Tax restrictions on internal entertainment. That can be unfortunate because in a global company, the diverse and separate sections that would benefit from the positive team building effect inherent to casual contact can be lost. At times, I pick up the bill personally and hope the gesture will be copied by others. That is also how I learned to pick up the breakfast bill first;)

During several supplier ARMS' Workshops in Europe with nationals from three or four countries participating. ARMS' meetings and other meetings of this type are conducted in English as the common language understood by all members. This happens even when there are no English or North Americans attending. English is the common engineering language understood across most western world countries.

Chapter Six

Risk Tree Analysis (RTA)

- Turning RTA on and building the Design Items report at the end of the ARMS' Workshop permits another method of reviewing the work accomplished.

- Facilitating Outstanding Leadership / Teamwork between all departments

- Engineering Design & Innovation Under Control

ARMS has the ability to calculate the assigned IPT (Incidents per Thousand) for each element of the design based on the relative Risks and IPT Start Value **(Fig.6-1)** assigned during the ARMS'

ITEM	Solid Oxide Fuel Cell (SOFC) SYSTEM / Next ARMS Review date ?	Workshop Date >>>> D.O.E.-R.E. / TRIZ	FEA / GOLDFIRE INNOV.	RED X / ENG. STUDY	Alpha / a	Support / Factor	GEN 1 λᵢ W/O (SF)	λᵢ W (-SF)
11	HOT ZONE MODULE (HZM)	55.0%			1.000006667		26.22	36.51
12	N-CELL STACK MODULE	100%	60%				14.36	22.30
13			100%					
14	GENERATES 42V POWER WITHIN STACK PERFORMANCE SPECIFICATIONS	8.250		25%	0.45	-40%	3.43	5.36
15	MAINTAINS GAS SEAL INTEGRITYAND ELECTRICAL ISOLATION	11.550		35%	0.50	-50%	4.36	7.95
16	PROVIDES ELECTRICAL CONNECTIONS TO POWER CONDITIONER	1.650		5%	0.20		1.12	1.12
17	MAINTAINS CELL (PEN) INTEGRITY	6.600		20%	0.40	-30%	3.03	4.10
18	PROVIDES STACK MOUNTING TO ICM	1.650		5%	0.30		0.92	0.92
19	PROVIDES ANODE OXIDATION PROTECTION	3.300		10%	0.40	-75%	1.51	2.85
20	FUEL REFORMER	20%					5.70	6.93
21			100%					
22	PROVIDES CONDITIONED AIR FOR FUEL MIXTURE PREPARATION	0.220		2%	0.20	-10%	0.15	0.16
23	PROVIDES FULLY VAPORIZED HOMOGENEOUS AIR/FUEL/ANODE GAS MIXTURE TO RE	1.980		18%	0.35	-20%	1.00	1.20
24	REACTOR/CATALYST PROVIDES REQUIRED REFORMATE QUALITY AND POWER OVER O	4.950		45%	0.40	-20%	2.27	2.81
25	GPC PROVIDES HEAT SOURCE TO CATHODE AIR PREHEAT (HEX) H2	1.100		10%	0.20	-10%	0.74	0.78
26	GPC REDUCES EXHAUST EMISSIONS AT OR BELOW REQUIRED LEVELS	1.650		15%	0.30	-50%	0.92	1.28
27	MAINTAINS GAS SEAL INTEGRITY	1.100		10%	0.30	-20%	0.61	0.71

Fig. 6-1

Workshop. These values are not percentages, they are increments of failure assigned from the Start Value. One way to use these values is as a pseudo-Risk Tree Analysis of the design, as perceived by the engineers. The RTA On and RTA Off buttons will make these calculations visible or not. The RTA result is similar to, but not a true, Fault Tree Analysis (FTA). The normal condition for this information should be hidden **(Fig.6-2)** to avoid confusion during the

Solid Oxide Fuel Cell (SOFC) SYSTEM / Next ARMS Review date ?	Workshop Date >>>> D.O.E.-R.E. / TRIZ	FEA / GOLDFIRE INNOV.	RED X / ENG. STUDY	Alpha / a	Support Factor
HOT ZONE MODULE (HZM)	55.0%			1.000006667	
N-CELL STACK MODULE	100%	60%			
		100%			
GENERATES 42V POWER WITHIN STACK PERFORMANCE SPECIFICATIONS			25%	0.45	-40%
MAINTAINS GAS SEAL INTEGRITYAND ELECTRICAL ISOLATION			35%	0.50	-50%
PROVIDES ELECTRICAL CONNECTIONS TO POWER CONDITIONER			5%	0.20	
MAINTAINS CELL (PEN) INTEGRITY			20%	0.40	-30%
PROVIDES STACK MOUNTING TO ICM			5%	0.30	
PROVIDES ANODE OXIDATION PROTECTION			10%	0.40	-75%
FUEL REFORMER		20%			
		100%			
PROVIDES CONDITIONED AIR FOR FUEL MIXTURE PREPARATION			2%	0.20	-10%
PROVIDES FULLY VAPORIZED HOMOGENEOUS AIR/FUEL/ANODE GAS MIXTURE TO RE			18%	0.35	-20%
REACTOR/CATALYST PROVIDES REQUIRED REFORMATE QUALITY AND POWER OVER O			45%	0.40	-20%
GPC PROVIDES HEAT SOURCE TO CATHODE AIR PREHEAT (HEX) H2			10%	0.20	-10%
GPC REDUCES EXHAUST EMISSIONS AT OR BELOW REQUIRED LEVELS			15%	0.30	-50%
MAINTAINS GAS SEAL INTEGRITY			10%	0.30	-20%

Fig. 6-2

ITEM	DESIGN ITEMS	INITIAL IPT	ALPHA	SF
15	MAINTAINS GAS SEAL INTEGRITYAND ELECTRICAL ISOLATION	11.550	0.50	-50%
14	GENERATES 42V POWER WITHIN STACK PERFORMANCE SPECIFICATIONS	8.250	0.45	-40%
17	MAINTAINS CELL (PEN) INTEGRITY	6.600	0.40	-30%
24	REACTOR/CATALYST PROVIDES REQUIRED REFORMATE QUALITY AND POWER OVER O	4.950	0.40	-20%
31	MAINTAINS GAS SEAL INTEGRITY	3.860	0.30	-10%
58	PROVIDES REGULATED AIR FLOW TO CIRCUITS	3.500	0.25	-30%
19	PROVIDES ANODE OXIDATION PROTECTION	3.300	0.40	-75%
55	MAINTAINS REQUIRED AIR FLOW & PRESSURE IN THE AIR VALVE MANIFOLD	3.000	0.40	-10%
195	MAINTAINS SENSOR ACCURACY WITHIN LIMITS	2.560	0.30	-30%
63	PROVIDES REQUIRED ANODE RECYCLE GAS FLOW TO THE AIR/FUEL VAPORIZER	2.400	0.40	-100%
57	OPERATES WITHIN NOISE EMISSION REQUIREMENTS	2.000	0.20	-50%
23	PROVIDES FULLY VAPORIZED HOMOGENEOUS AIR/FUEL/ANODE GAS MIXTURE TO RE	1.980	0.35	-20%
204	MAINTAINS SENSOR ACCURACY WITHIN LIMITS	1.920	0.30	-30%
16	PROVIDES ELECTRICAL CONNECTIONS TO POWER CONDITIONER	1.650	0.20	
18	PROVIDES STACK MOUNTING TO ICM	1.650	0.30	
26	GPC REDUCES EXHAUST EMISSIONS AT OR BELOW REQUIRED LEVELS	1.650	0.30	-50%
30	PROVIDES HEAT TRANSFER FROM EXHAUST TO CATHODE AIR	1.650	0.30	-10%
74	MAINTAINS ELECTRICAL CONNECTIONS	1.650	0.10	-30%

Fig. 6-3

ARMS' Workshop and only be reviewed with the team after the ARMS' model is complete. The relative RTA values should agree with the team's perception of the design. Keep in mind RTA values will shift up or down when the start value failure rate is calibrated. The RTA values are used by ARMS to Pareto rank the design items in the "Design Items" report **(Fig.6-3)**.

Turning RTA on and building the Design Items report at the end of the ARMS' Workshop permits another method of reviewing the work accomplished.

If the order presented in the Design Items report does not appear to be correct to the team, there is more work to be done to balance out the % risk values assigned. Usually a few adjustments need to be made, because the Design Items report looks at the whole design when forming ranks, and it can bring errors to the top of the list or show that an important item has not been given enough attention. The Design Items report will also bring more attention to any negative Support Factors that end up near the top of the list.

There is a synergy between information developed in an ARMS' Workshop and a FMECA development that speeds up the FMECA process greatly. ARMS and FMECA are different in a major sense.

FMECA requires the identification of specific failure modes. To do so requires significant information about the design only available in the latter phases of the design activity. ARMS can start at the conception stage and provide support throughout the design process.

Chapter Seven

Failure Analysis Reporting (FAR)

- ARMS supports the use of small sample sizes to project on a Weibull graph how the design is progressing.

- It is important to spend money on obtaining samples with dimensions at nominal. You will save much greater sums of money in the future on this and future projects.

- Early in the development, it is more efficient to use your limited budget completing Assignable Actions and solving new problems revealed by Failure Analysis.

- Facilitating Outstanding Leadership / Teamwork between all departments

- Engineering Design & Innovation Under Control

There are a number of excellent Failure Analysis Report programs available through the Internet. It is not the purpose of this section to tell you which software to use, how to conduct a failure analysis or what to analyze or report. The real question is, why conduct a failure analysis and when, particularly in an ARMS environment?

When engineering a new product, be it electronic, mechanical, chemical or a hybrid combination, it usually is expensive to obtain prototypes for durability testing. This is true for "one off" products like aerospace applications, mid-volume products like HVAC compressors (millions a year) or high-volume vacuum cleaner motors (millions a month). In the first case, high cost per sample is obvious, in the high-volume case it is because you need high volume production runs to obtain samples representative of production intended configurations. The production process becomes a part of the design. If you need a few to test, you first need to build thousands or hundreds of thousands to have representative units.

ARMS supports the use of small sample sizes to project on a Weibull graph how the design is progressing.

A n/(n-1) bias for small sample sizes makes adjustments to the

PEI 7-1-W[1]			1 Γ =		PROJECT NAME?				
TOTAL 7			1	< Acceleration Factor	Line color 255, 0, 0	Beta Est.1 >>	0.60	1	1.30E+08
Line #	QTY TAKEN OFF TEST	Failure=1 Suspension=0	Rank Individual TIMES or CYCLES at Test Termination	INCREMENT	Mean Order Number	NIBPSS	Median Rank %	Plot Point Y	Plot Point X
1			CYCLES		1.00				
1	1	1	6,800,000	1.0000	1.000		9.428%	-2.3124	6.80E+06
2	1	1	10,200,000	1.0000	2.000		22.952%	-1.3442	1.02E+07
3	1	1	15,000,000	1.0000	3.000		36.476%	-0.7902	1.50E+07
4	1	1	19,000,000	1.0000	4.000		50.000%	-0.3665	1.90E+07
5	1	1	32,000,000	1.0000	5.000		63.524%	0.0085	3.20E+07
6	1	1	65,000,000	1.0000	6.000		77.048%	0.3865	6.50E+07
7	1	1	130,000,000	1.0000	7.000		90.572%	0.8593	1.30E+08
8									

Fig. 7-1

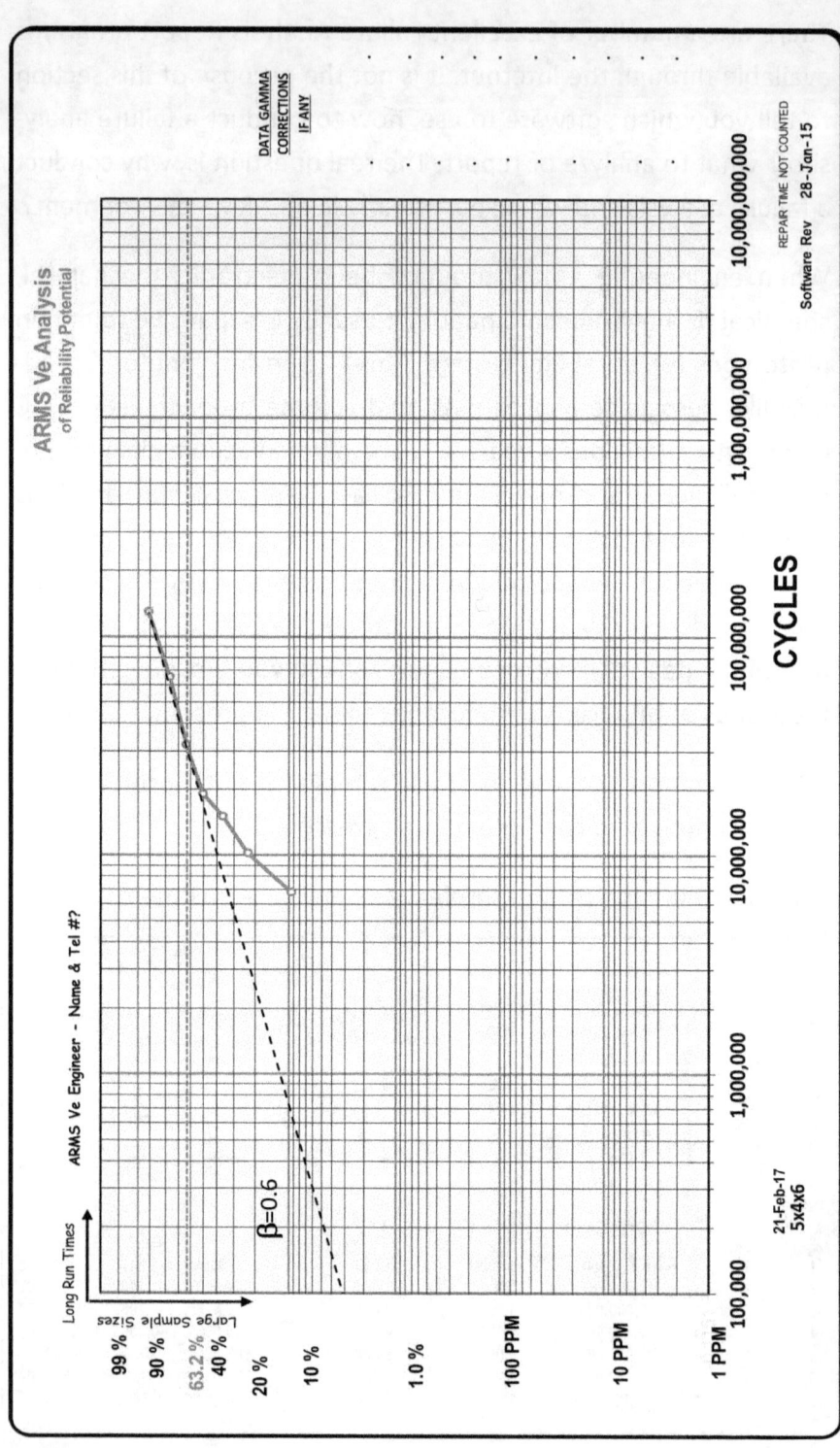

Fig. 7-2

data.

To illustrate how this is accomplished, durability tests on small batches of piezo electric fuel injectors (PEI) were made as the design was being improved. On the first data set **(Fig.7-1),** 8 PEI were durability tested. One was taken off test early for analysis at 10,000 injections and the remaining 7 were run to failure without being removed from the test stand. The result was a Weibull slope of approximately 0.6 **(Fig. 7-2)**. The single injector taken off test early without failure was removed from the test data. It did not run or fail and was making no contribution to the analysis, therefore 7/7 failures were reported.

This suggests the injectors are in a premature failure region of a bathtub curve. The slope is represented by the dashed vector passing through the last 3 of 7 failure points. These results are troubling. It would appear there are clear defects in the first 4 injectors and they may not be representative of the balance of the lot. When a Weibull graph is convex at the start of the trace, it usually is a sign of bi-modal failure causes. Failure analysis was required to discern what was occurring. The test results plotted on a Weibull graph are used to direct the attention of the Failure Analyst actions. Failure analysis must also include review of all build data, signals data during durability test at the start and prior to failure, physical analysis of failed units and any surviving units. The Failure Analyst must be supported with the availability of equipment or services that would shed light on the cause of failure. Often, there are abnormal conditions just before failure that can be utilized to catch a failure in the earliest stages and make the failure analysis job a lot easier. It may be a slight noise, strange electrical wave form, radiation, vibration, current surge or other outward sign. The Failure Analyst may be able to equip the test stands with an ability to sense the change and shut the test down. The Failure Analyst needs to be supported in this effort.

Once a failure occurs, the hardware needs to be totally under the control of the Analyst. Treat the failure like a crime scene.

In this case, there were physical phenomena suspected of causing the first four units to fail. The batch durability test had confirmed bi-modal failure modes. The convex hump in the data, as suspected, is from bi-modal failures. We are looking at the combination of at least two bath tubs in the data.

This data was further analyzed by decomposing the data into two sets, labeled **[2]** and [3] **(Fig.7-3)**. (The first four injectors and the last three.)

ARMS' Weibull **(Fig. 7-4)** indicates the split with 4 injectors, labeled **[2]**, above the original line and 3 injectors, labeled [3], below. This can be interpreted as one failure mode causing early failures and the other causing failures after greater exposure. This knowledge can assist the Failure Analyst in looking for the differ-

PEI 7-3 DC [2]			2 Γ =		PROJECT NAME?					
TOTAL	4		1.0	< Acceleration Factor	Line color 0, 0, 0	Beta Est.1 >>	2.30	1	19,000,000
Line #	QTY TAKEN OFF TEST	Failure=1 Suspension=0	Rank Individual TIMES or CYCLES at Test Termination	INCREMENT	Mean Order Number	NIBPSS	Median Rank %	Plot Point Y	Plot Point X
2			CYCLES		1.00				
1	1	1	6,800,000	1.0000	1.000		15.910%	-1.7528	6,800,000
2	1	1	10,200,000	1.0000	2.000		38.637%	-0.7167	10,200,000
3	1	1	15,000,000	1.0000	3.000		61.363%	-0.0503	15,000,000
4	1	1	19,000,000	1.0000	4.000		84.090%	0.6088	19,000,000
5									

PEI 7-1 DC [3]			3 Γ =		PROJECT NAME?					
TOTAL	3		1.0	< Acceleration Factor	Line color 0, 155, 0	Beta Est.1 >>	1.00	1	130,000,000
Line #	QTY TAKEN OFF TEST	Failure=1 Suspension=0	Rank Individual TIMES or CYCLES at Test Termination	INCREMENT	Mean Order Number	NIBPSS	Median Rank %	Plot Point Y	Plot Point X
3			CYCLES		1.00				
1	1	1	32,000,000	1.0000	1.000		20.630%	-1.4651	32,000,000
2	1	1	65,000,000	1.0000	2.000		50.000%	-0.3665	65,000,000
3	1	1	130,000,000	1.0000	3.000		79.370%	0.4564	130,000,000
4									

Fig.7-3

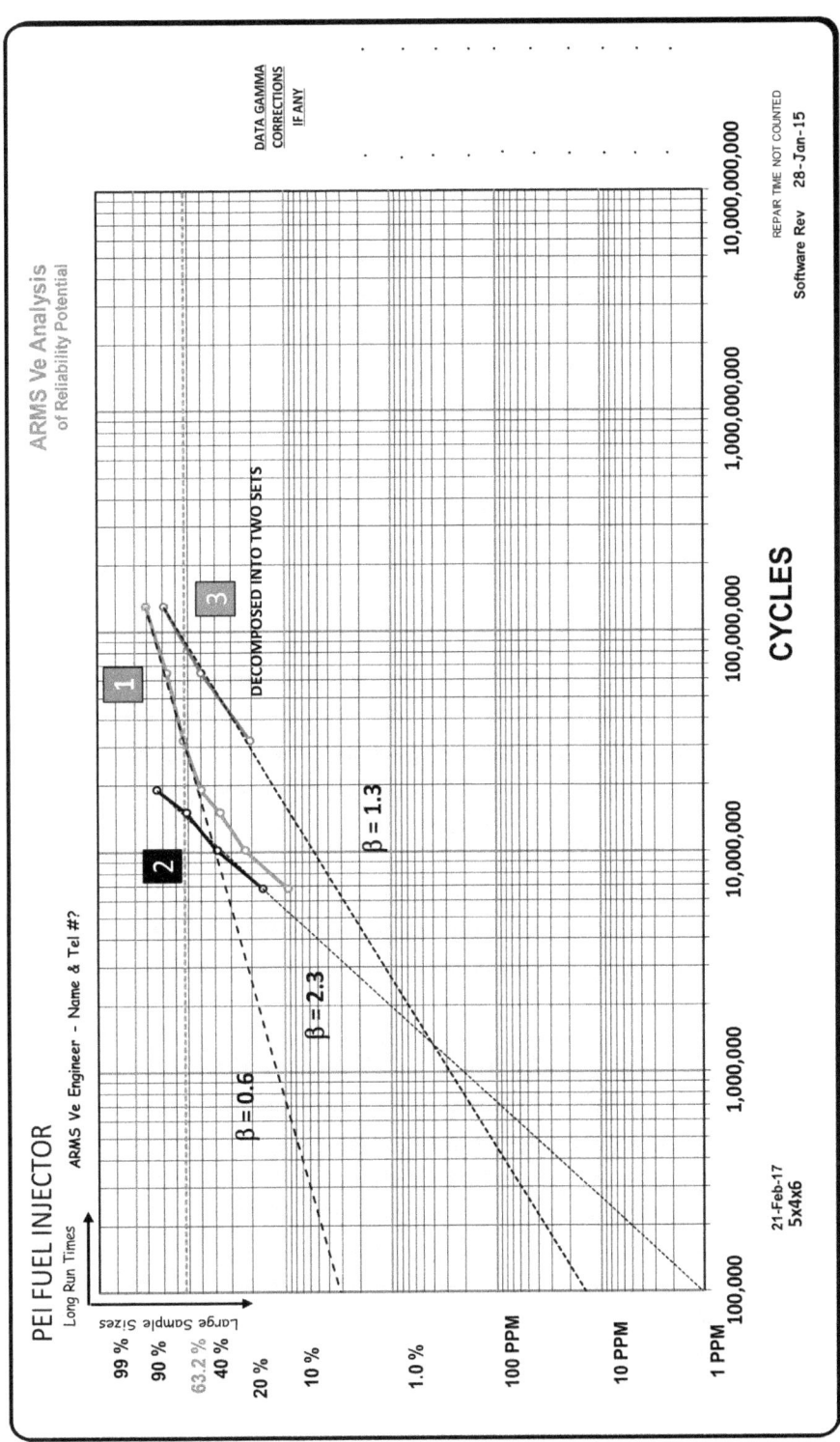

Fig. 7-4

143

ence in the two sets that was apparent in the data. In this case, the differences in the hardware were not easily detected and may have been overlooked without the heads-up input from data decomposition.

The ARMS' Weibull of the decomposed data **(Fig. 7-4)** exhibited β slopes of 2.3 and 1.3. The four earlier failures **[2]** were due to identified defects in the assembly. The wearout-type failure mechanism [2] turned out to be a PEI sealing failure. The last three [3] failed catastrophically with a great deal of material being "carburized." The cause of failure was not determined beyond it not being the seal failure of the first four. The β slope of 1.3 suggests the failures were random and from something other than a classical wearout condition.

After seeing the degree of "carburization" taking place during failures, new circuits were added to the test stands to limit current flow during failure events and to shut the test down. The intention was to minimize the destructive failure conditions. Analysis of failures now includes; the failure of test stands to protect conditions at the start of failure. Energy within the test system contained in capacitors, inductors, chemicals or heat exchangers can cause catastrophic failure without an external ability to sense the initiation of failure. Lithium Batteries, Piezo Electric Injectors, SOFC, Fly Wheel Propulsion, Jet Engines and Sterling Thermal Engines all have this test set-up Assignable Action to be solved. If energy resides in the operating system, it must be dealt with when you stop. Monitoring electromagnetic fields can be helpful. Another often missed failure cause is 3 phase motors being powered by an unbalanced power source. There was a time when this was a problem in downtown Houston, Texas, due to a "Dog Leg Ground" power distribution system.

During an ARMS' Workshop, and while centered on durability

test and failure analysis results, the team decided to make a "Queen of Hearts" change (if you can't fix it, throw it out) in the design that would abandon certain concepts of the design and switch to a new PEI sealing approach, hoping to improve results. This involved throwing out a section of the engineering build equipment and replacing it with more qualified production equipment that was built for the new PEI sealing element design. This was not a total change but early deletion of a temporary approach that was never intended for production, but employed while awaiting for   delivery of production grade equipment. The change also involved electrical connection improvements.

After incorporation of Assignable Actions in the design and build processes, a second batch of 9 PEI components were made and subjected to durability testing. Similar results were obtained, 6/9 failed (Fig.7-5), however, at a greater number of cycles in the test program. The longest running three have a slope of β=0.4. Reviewing the Weibull graph of all failures (Fig. 7-6) reveals three failures in a cluster near 100,000,000 cycles. In other words, the batch is again bi-modal. The best way to determine what is happening is through failure analysis directed by data decomposition as shown in the following data tables (Fig. 7-7 & Fig.7-8) and

PEI 7-3[4]			4 Γ =		PROJECT NAME?					
TOTAL	9		1.0	< Acceleration Factor	Line color 200, 100, 0	Beta Est.1 >>	0.40	1	5.7E+08	
Line #	QTY TAKEN OFF TEST	Failure=1 Suspension=0	Rank Individual TIMES or CYCLES at Test Termination	INCREMENT	Mean Order Number	NIBPSS	Median Rank %	Plot Point Y	Plot Point X	
4			CYCLES		1.00					
1	1	1	4.8E+07	1.0000	1.000		7.413%	-2.5637	4.8E+07	
2	1	1	9.7E+07	1.0000	2.000		18.059%	-1.6136	9.7E+07	
3	1	1	1.1E+08	1.0000	3.000		28.706%	-1.0836	1.1E+08	
4	1	1	1.2E+08	1.0000	4.000		39.353%	-0.6929	1.2E+08	
5	1	1	2.7E+08	1.0000	5.000		50.000%	-0.3665	2.7E+08	
6	1	1	5.7E+08	1.0000	6.000		60.647%	-0.0698	5.7E+08	
7	3							-0.0698	5.7E+08	
8										

Fig. 7-5

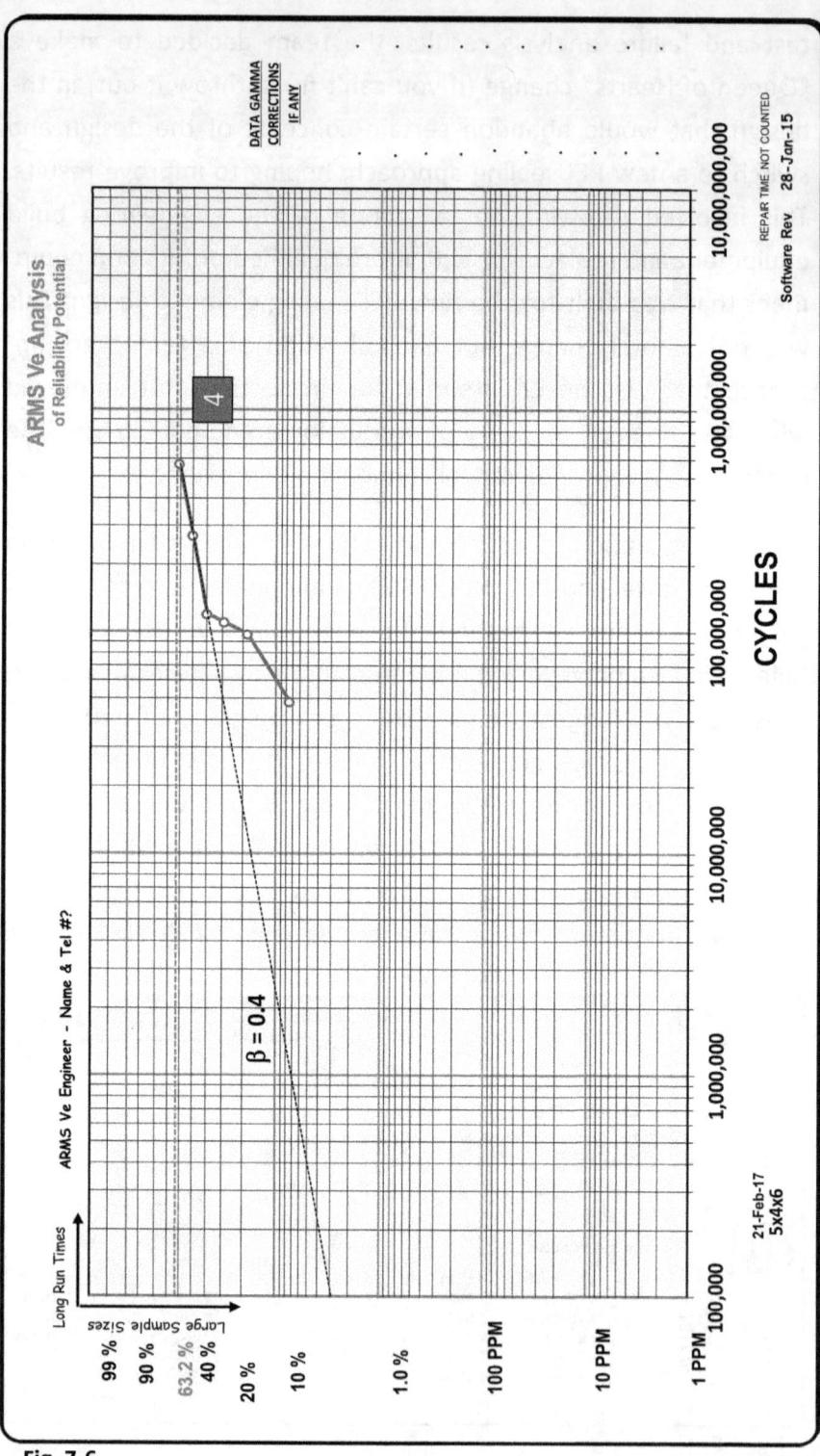

Fig. 7-6

146

Weibull graph **(Fig. 7-9)**. In this case two failure modes were revealed and corrections applied. Additional examples of **GUI** decomposed data will be covered later in Chapter 12.

Production personnel at the ARMS' Workshop stepped up and indicated that they would support the decision to secure production grade equipment to replace the engineering prototype equipment. Their support, as team members, was important in obtaining the Production Dept. approval. It was believed the early results would improve radically. To make this equipment available required capital and other management adjustments. The necessary equipment existed within the company, but was controlled by the production department. Operating time on the equipment was not easily obtained. After engineering development managers negotiated with production managers, the equipment issue was resolved.

PEI 7-2-DD[5]			5 Γ =		PROJECT NAME?					
TOTAL	8		1.0		< Acceleration Factor	Line color 0, 175, 250	Beta Est.1 >>	6.00	1	1.2E+08
Line #	QTY TAKEN OFF TEST	Failure=1 Suspension=0	Rank Individual TIMES or CYCLES at Test Termination	INCREMENT	Mean Order Number	NIBPSS	Median Rank %	Plot Point Y	Plot Point X
5			CYCLES		1.00				
1	1	1	97,000,000	1.0000	1.000		8.300%	-2.4460	9.7E+07
2	1	1	110,000,000	1.0000	2.000		20.214%	-1.4880	1.1E+08
3	1	1	120,000,000	1.0000	3.000		32.128%	-0.9479	1.2E+08
4	5							-0.9479	1.2E+08
5									

Fig. 7-7

PEI 7-1 DC [3]			3 Γ =		PROJECT NAME?					
TOTAL	6		1.0		< Acceleration Factor	Line color 0, 155, 0	Beta Est.1 >>	1.30	1	130,000,000
Line #	QTY TAKEN OFF TEST	Failure=1 Suspension=0	Rank Individual TIMES or CYCLES at Test Termination	INCREMENT	Mean Order Number	NIBPSS	Median Rank %	Plot Point Y	Plot Point X
3			CYCLES		1.00				
1	1	1	32,000,000	1.0000	1.000		10.910%	-2.1583	32,000,000
2	1	1	65,000,000	1.0000	2.000		26.546%	-1.1760	65,000,000
3	1	1	130,000,000	1.0000	3.000		42.182%	-0.6017	130,000,000
4	3							-0.6017	130,000,000
5									

Fig. 7-8

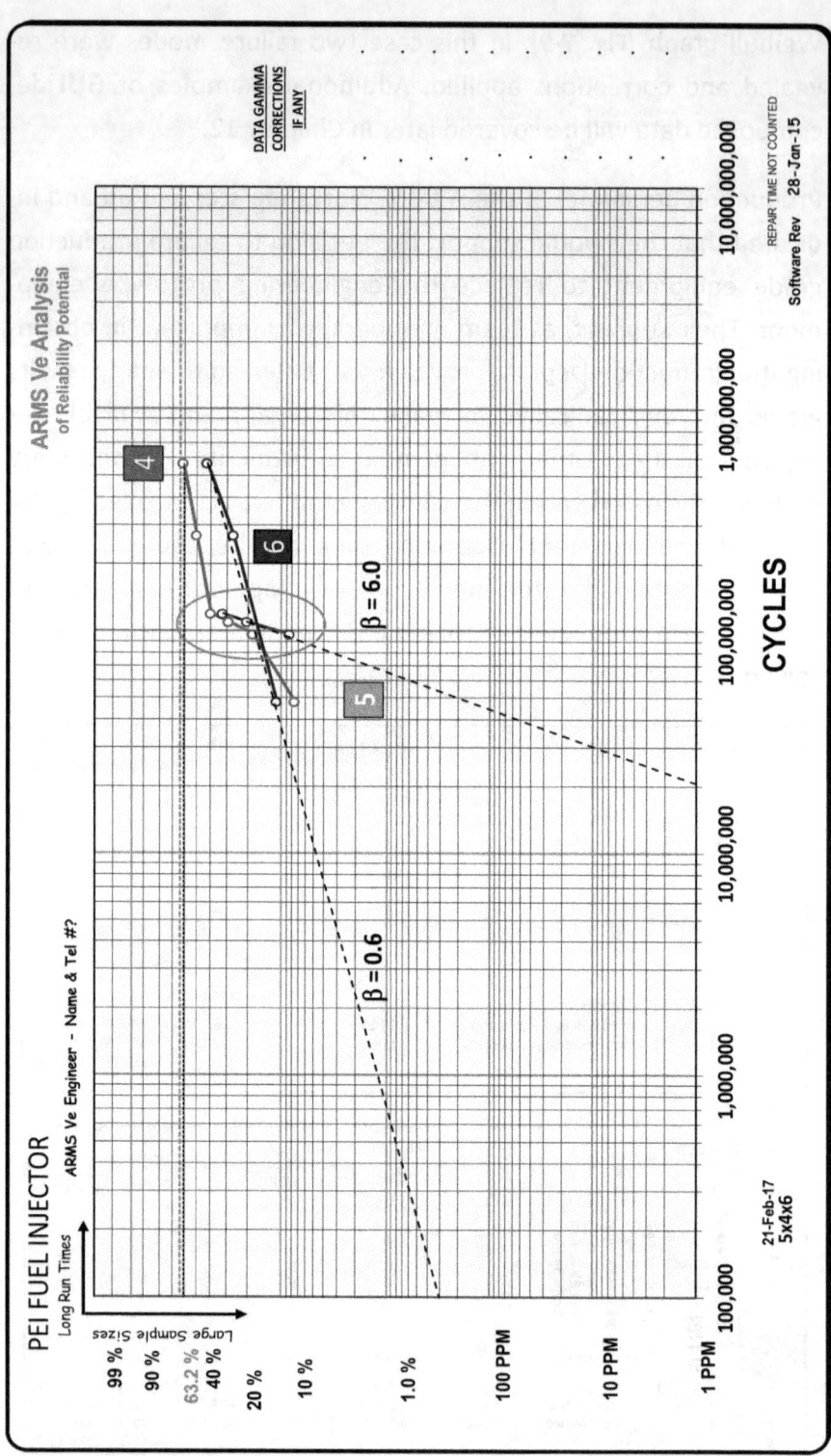

Fig. 7-9

148

The three PEI units clustered after the first failure appear to be "birds of a feather" **(Fig. 7-6)** and are treated as such in the Weibull analysis. The first failure is placed in the same set as the last two **(Fig. 7-9)**.

The next test batch validation tested were 3 survivors from the 6/9 test **(Fig. 7-5)**. The 3 PEI that had passed up to that number of cycles were returned to the test stands. They yielded the data results of **(Fig. 7-10)** and the Weibull analysis as shown on **(Fig. 7-11)**.

ARMS has an additional tool which will apply +/- 1.645σ to the Median. This produces a +/- 95% range limit. For our sample of three, they can be seen on Weibull graph **(Fig. 7-11)** circled in red. There appears to be an improvement, but the wide median range is not very comforting. The Median range values are good for visually indicating the problem of working with small sample sizes, particularly if the %σ of the build process is not tight.

It is important to spend money on obtaining samples with dimensions at nominal. You will save much greater sums of money in the future on this and future projects.

Results from durability testing 24 additional PEI units are shown in the data set of **(Fig. 7-12)** and Weibull graph **(Fig. 7-13)**. Clearly a breakthrough has been achieved. There are no premature failures or bimodal patterns and the wearout region has shifted far to the right. This result may or may not be sufficient. It depends

Line #	QTY TAKEN OFF TEST	Failure=1 Suspension=0	Rank Individual TIMES or CYCLES at Test Termination	INCREMENT	Mean Order Number	NIBPSS	Median Rank %	Plot Point Y	Plot Point X	"Applied Life Data Analysis" Wayne Nelson	"Applied Life Data Analysis" Wayne Nelson
PEI 7-2-DD[7]		7 Γ =	5.7E+08	PROJECT NAME?						Sample Size 95% Conf (+)	Sample Size 95% Conf (-)
TOTAL 3		1.0	< Acceleration Factor	Line color 140, 210, 80	Beta Est.1 >>	10.00	1	8.9E+08	1	1
			CYCLES		1.00					+ θ	− θ
1	1	1	180,000,000	1.0000	1.000		20.630%	-1.4651	7.5E+08	1,938,780,409	290,130,846
2	1	1	240,000,000	1.0000	2.000		50.000%	-0.3665	8.1E+08	2,093,882,842	313,341,313
3	1	1	320,000,000	1.0000	3.000		79.370%	0.4564	8.9E+08	2,300,686,086	344,288,604
4											

Fig. 7-10

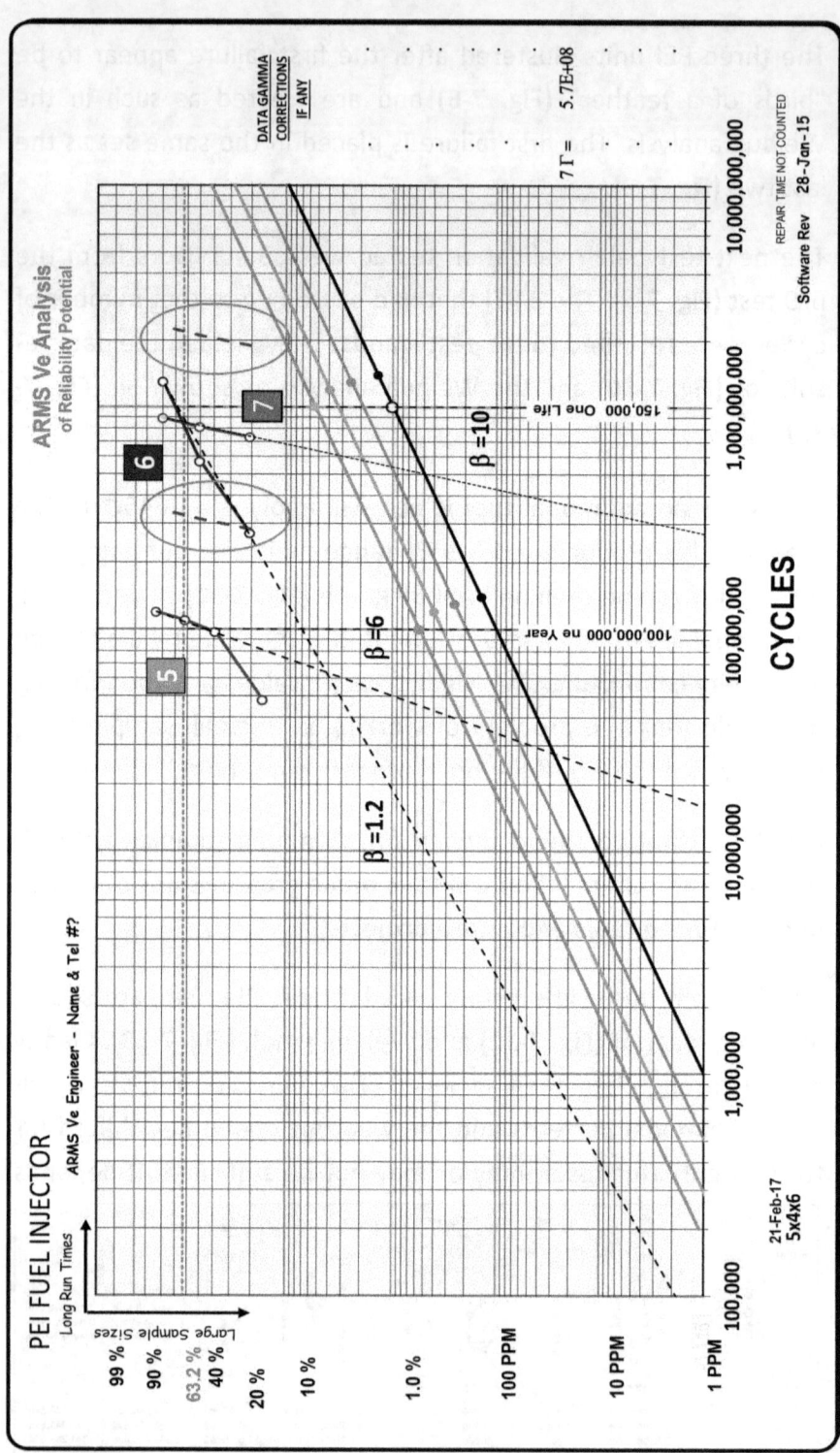

Fig. 7-11

150

on the goals that we have been asked to meet.

As in the previous graph, a +/-95% approximation of a median range limit has been added by opening the Validation Distance tab and entering a 1 & 1 in the appropriate Data Entry sheet. The tighter +/-95% limits displayed are due to the larger sample size of 24. It makes sense to work with small sample sizes and make fast design turns while the design is still fluid and later increase the sample sizes when sample costs are lowered. Larger sample sizes confirm your ability to make units consistently as well as the integrity of the design.

Early in the development, it is more efficient to use your limited budget completing Assignable Actions and solving new problems revealed by Failure Analysis.

We have to reap the benefits from our work on Assignable Actions and the Failure Analysis. Before spending more budget on making test samples, make sure the Assignable Actions and Failure Analysis recommendations are being completed and fully integrated into the next design and build cycle. This is not a time to take short cuts on solution integration. In this case, when the additional as-

PEI 7-3-DATA[8]		8 Γ =		PROJECT NAME?						Sample Size 95% Conf (+)	Sample Size 95% Conf (-)
TOTAL	24	1	< Acceleration Factor	Line color 0, 100, 200	Beta Est.1 >>	7.00	1	910,000,000		1	1
Line #	QTY TAKEN OFF TEST	Failure=1 Suspension=0	Rank Individual TIMES or CYCLES at Test Termination	INCREMENT	Mean Order Number	NIBPSS	Median Rank %	Plot Point Y	Plot Point X	"Applied Life Data Analysis" Wayne Nelson	"Applied Life Data Analysis" Wayne Nelson
8			CYCLES		1.00					$+\theta$	$-\theta$
1	1	1	520,000,000	1.0000	1.000		2.847%	-3.5446	5.20E+08	7.27E+08	3.72E+08
2	1	1	620,000,000	1.0000	2.000		6.947%	-2.6311	6.20E+08	8.67E+08	4.43E+08
3	1	1	650,000,000	1.0000	3.000		11.047%	-2.1450	6.50E+08	9.09E+08	4.65E+08
4	1	1	670,000,000	1.0000	4.000		15.148%	-1.8063	6.70E+08	9.37E+08	4.79E+08
5	1	1	690,000,000	1.0000	5.000		19.248%	-1.5428	6.90E+08	9.65E+08	4.93E+08
6	1	1	710,000,000	1.0000	6.000		23.348%	-1.3246	7.10E+08	9.93E+08	5.07E+08
7	1	1	720,000,000	1.0000	7.000		27.448%	-1.1367	7.20E+08	1.01E+09	5.15E+08
8	1	1	730,000,000	1.0000	8.000		31.549%	-0.9701	7.30E+08	1.02E+09	5.22E+08
9	1	1	740,000,000	1.0000	9.000		35.649%	-0.8191	7.40E+08	1.04E+09	5.29E+08
10	1	1	760,000,000	1.0000	10.000		39.749%	-0.6799	7.60E+08	1.06E+09	5.43E+08
11	1	1	770,000,000	1.0000	11.000		43.850%	-0.5497	7.70E+08	1.08E+09	5.50E+08
12	1	1	780,000,000	1.0000	12.000		47.950%	-0.4262	7.80E+08	1.09E+09	5.58E+08
13	1	1	820,000,000	1.0000	13.000		52.050%	-0.3079	8.20E+08	1.15E+09	5.86E+08
14	1	1	840,000,000	1.0000	14.000		56.150%	-0.1931	8.40E+08	1.18E+09	6.00E+08
15	1	1	850,000,000	1.0000	15.000		60.251%	-0.0806	8.50E+08	1.19E+09	6.08E+08
16	1	1	890,000,000	1.0000	16.000		64.351%	0.0310	8.90E+08	1.25E+09	6.36E+08
17	1	1	910,000,000	1.0000	17.000		68.451%	0.1429	9.10E+08	1.27E+09	6.50E+08
18	7							0.1429	9.10E+08	1.27E+09	6.50E+08
19											

Fig. 7-12

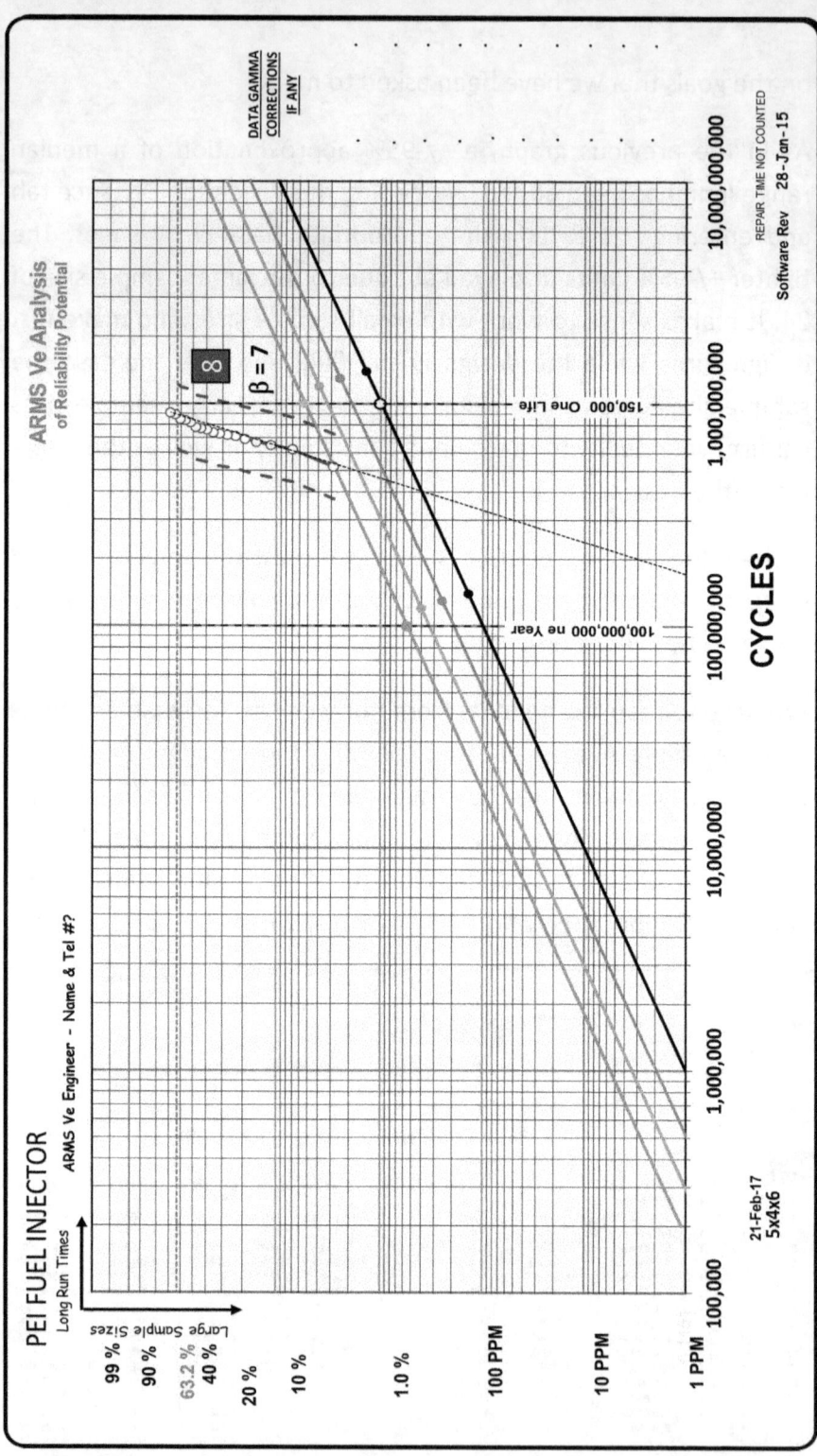

Fig. 7-13

152

signed work was completed, new hardware was built.

It is very important to the ARMS' model's contribution to make sure the line numbers affected by an action have the line number assigned to the action. This is difficult and always comes at a time the team is getting tired and ready to pass the task off to an individual. Persevere, get the job done right. Errors of omission can cause important actions to drop down on the Pareto ranking.

A "5000 foot" review of the "Design Items" and "Assignable Actions" lists, to see if they make sense to the team, can help determine if a more rigorous review of the ARMS' model is necessary. If you are assigned responsibility for an improvement action, you want to make sure the action is receiving a high risk number. High risk numbers secure resources and increase the probability of obtaining solutions. Risk assignment is a competitive process. Obtaining risk assignments is the "get your piece of the pie" idea. It is not a time to be out of the room. When you review the Design Items and Assignable Actions reports, make sure you scroll down to the bottom looking for design items or assignable actions that were not assigned values or did not receive all of the assignments that they should have.

Chapter Eight

ARMS Weibull Analysis,

Failure Rate,

Reliability Growth,

MTTF

- The same source data is utilized to express Failure Rate, Relibility Growth, MTTF and Weibull graphs.

- Facilitating Outstanding Leadership / Teamwork between all departments

- Engineering Design & Innovation Under Control

Different industries tend to express risk reduction or reliability improvement utilizing one of a few graphical methods. During ARMS' meetings while we are viewing a Weibull graph, we hear things like, "What is that in terms of meantime to failure (MTTF)?" Or "What is the expected improvement in Reliability Growth (RG)?" Or "What will the improvement in Failure Rate (FR) be?" All of these outputs are different forms of representing the same raw data. With ARMS, we have elected to focus on a Weibull representation. Once you are familiar with Weibull Analysis techniques, they provide the most insight to risk information. To illustrate this, one set of data will be represented using each of the methods:

Weibull **(Fig. 8-1)**

FR   **(Fig. 8-2)**

RG   **(Fig. 8-3)**

MTTF   **(Fig. 8-4)**

Data was taken from an ARMS' model early in the development of a gasoline fuel injector. The data contains points for each of three forecasted generations of design. The end values have an additional point to display the effect of reported negative Support Factors.

1) FR = (# of failures) / (Total time of all units on test). = λ

2) Reliability = $(1 - \lambda)$ $\beta = 1$

3) MTTF = $(1 / \lambda)$ $\beta = 1$

4) MTTF $_{95\%Limits}$ = MTTF*$((1 +/- 1.96) *\sigma* n/(n-1))$

5) Failure Rate $_{95\%Limits}$ = λ+/- $(1.96 * n/(n-1))$

The SFs result in an adjustment step between each generation. The failure rate graph **(Fig. 8-2)** indicates the FR after the equivalent of one year in service + planned extended life for each of the three generations of design.

The initial START FR (10%) and the next three generations of design FR points are plotted and connected with an exponential line to form a smooth progression **(Fig. 8-2)**:

10% at one year 500 hours, Start Design, (100/1000)

6.8% at one-year Design Gen1 with SF applied,

4.7% at one-year Design Gen2 with SF applied,

2.8% at one-year Design Gen3 with SF applied.

At the end of one life time the % FR One Life, indicates:

65.1% at one life, 5000 hours, for the initial design,

50.8%, at one life, Design Gen 1 with SF applied

38.1%, at one life, Design Gen 2 with SF applied

25.0%, at one life, Design Gen 3 with SF applied

In this example one year of service was 500 Hours and one life of service was 10 years which equals 5000 Hours.

On the ARMS' Weibull graph **(Fig.8-1)** these same values of % Failure Rate are found at the 5000-hour vertical axis on the Weibull graph where lines representing the initial and Gen1-3 cross the vertical. These points are highlighted with a DOT on the ARMS' Weibull graph.

ARMS calculates the end points of the curves, then fills in an expo-

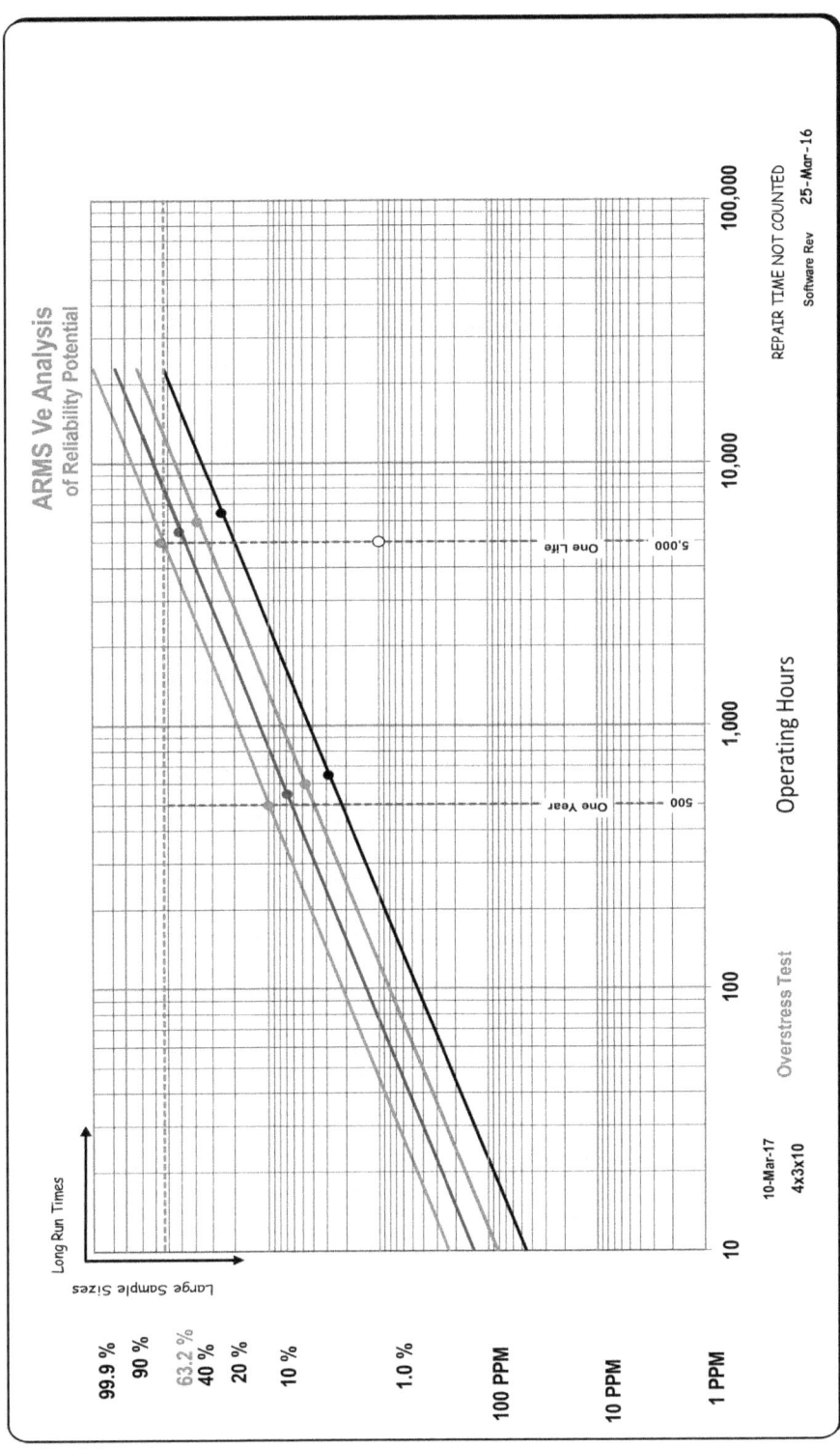

Fig. 8-1

157

nential connecting curve. The values in between the end points have no meaning and the end points could be connected by a straight line or any other curved line. We do not read or interpret between end points.

The Reliability Growth and Mean Time to Failure are calculated directly from the failure rate numbers assuming $\beta = 1$. Values at the reported $\beta <> 1$ would require additional calculations for each slope.

Again, four outputs for the same data set. ARMS concentrates on the Weibull representation **(Fig. 8-1)** because, once familiar, it is capable of conveying more information without additional calculations.

ARMS' Weibull graphs are also useful for reviewing different projects on one of three graphs. There are three different Weibull graphs in ARMS, FR % vs Time, FR % vs Distance and FR % vs Cycles. With a common goal, you can judge how close to the goal each project is and when they might reach their goal. If the ARMS' models are a group of sub-systems for a project, the Weibull will reveal where plans are falling behind. This is a visual GUI system overview that can make a good posting.

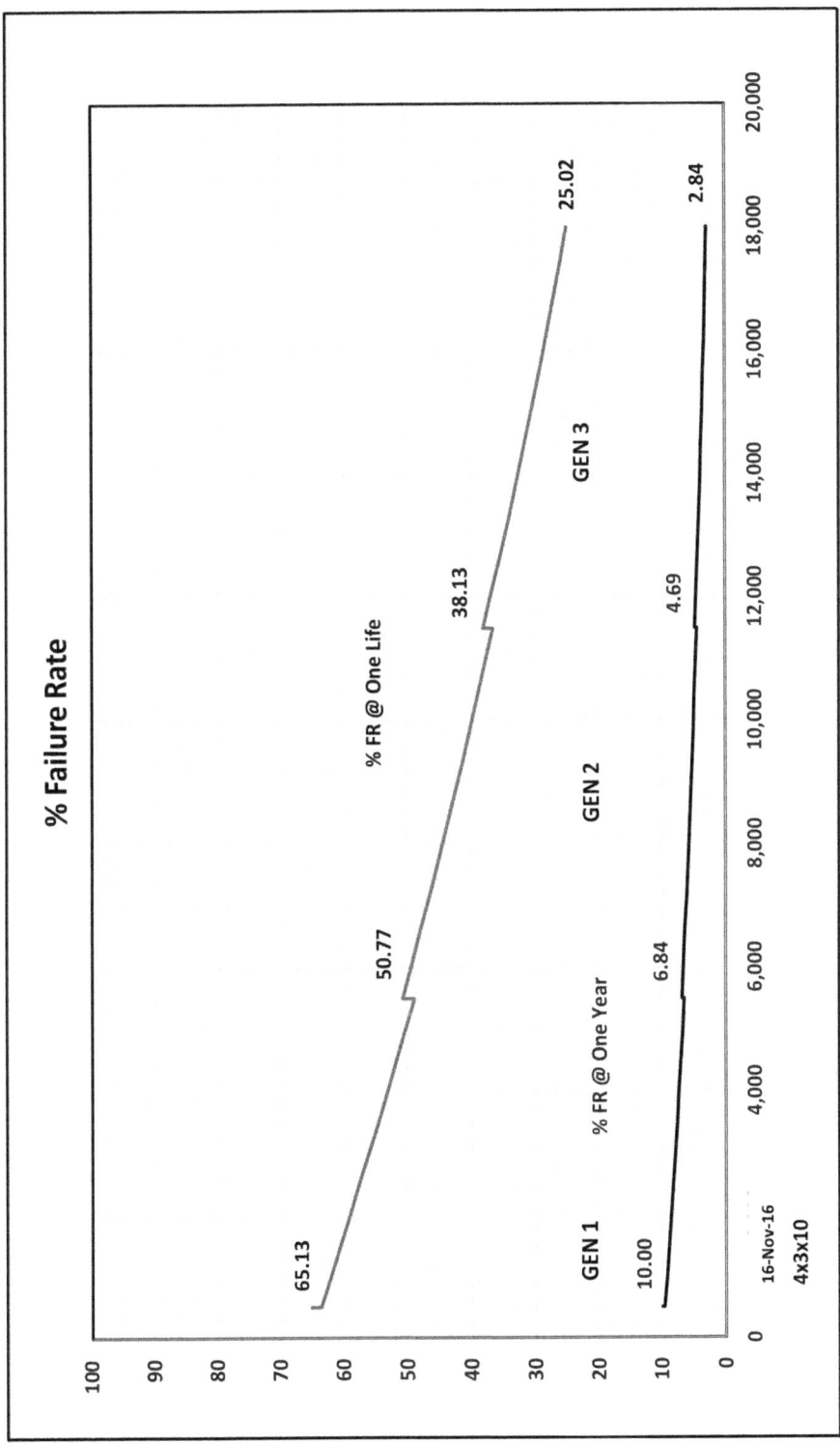

% Failure Rate

Fig. 8-2

159

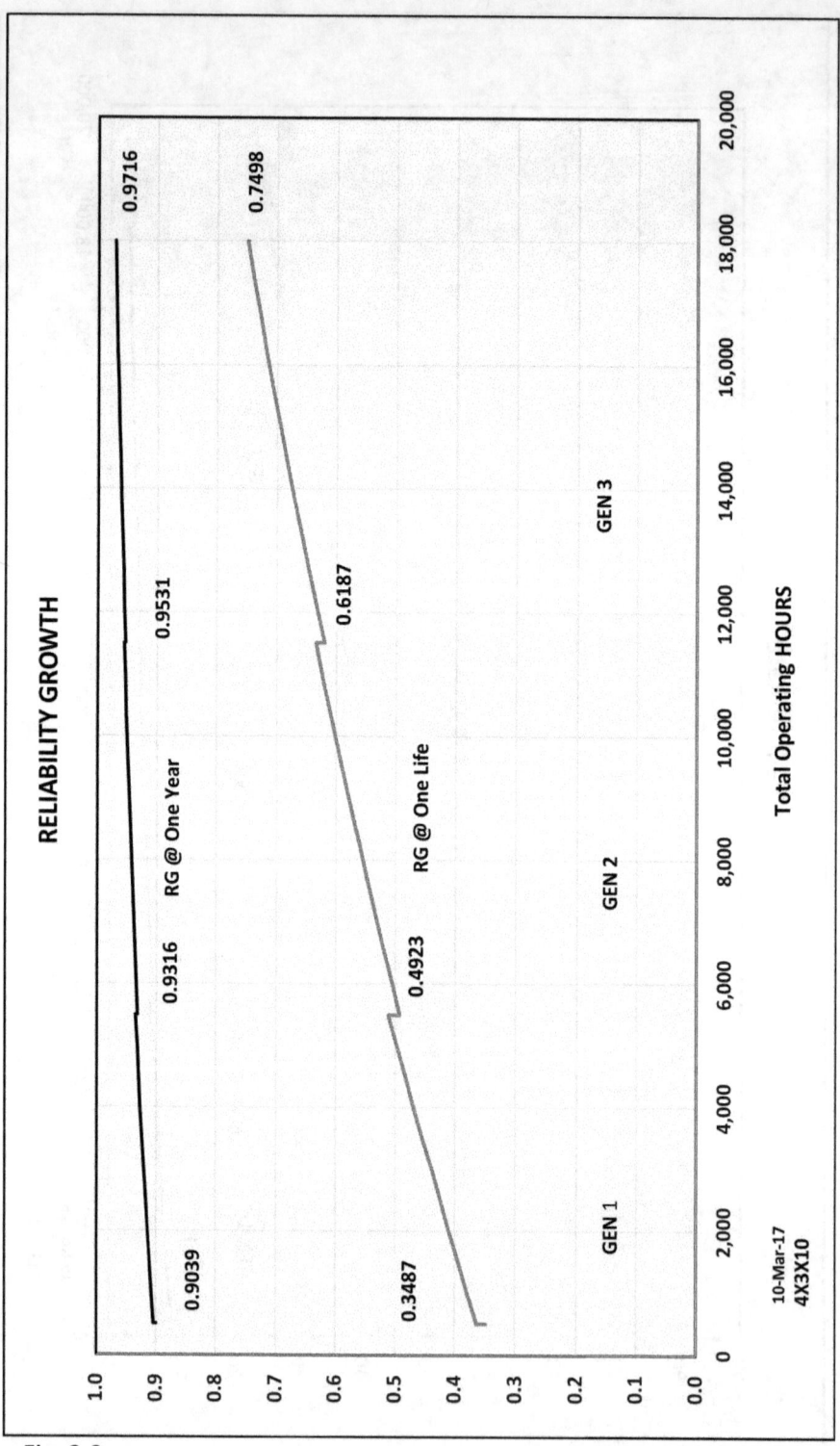

Fig. 8-3

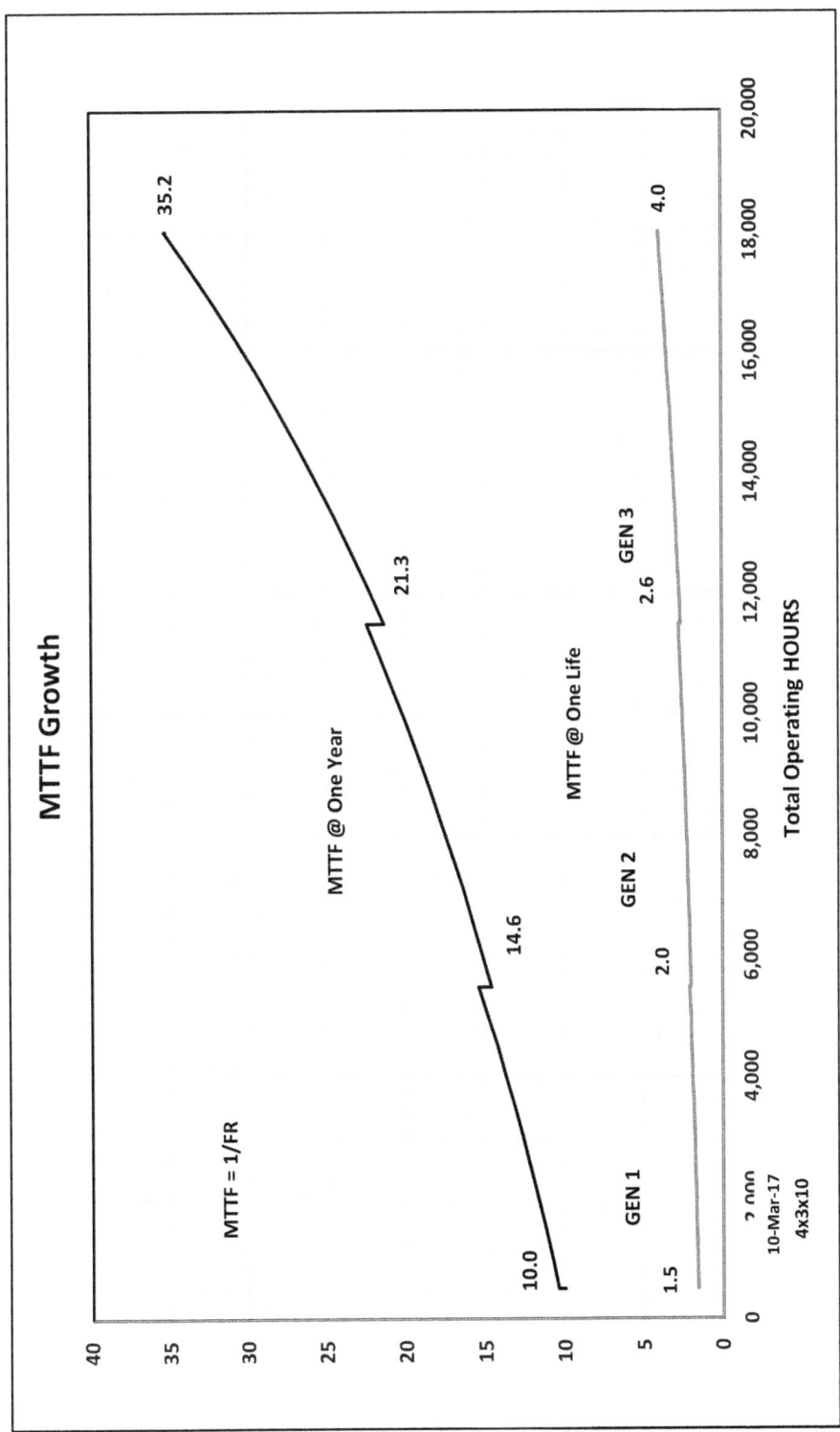

Fig. 8-4

161

We can examine the results from durability testing a sample of 24 electro-mechanical units.

After about 50 hours of testing, the units were taken off the durability stand and subjected to a series of tests to verify performance. The combined total of initial engineering acceptance test, 50 hours of durability test, and post durability test evaluations was not known accurately but estimated to be the equivalent of 80 hours of normal operation. After incorporating limited design improvements, the 24 units were put back on the durability test stand (80 hours = Γ 2400 miles) at a design specification average of 30 mph. The applied design improvements were within the Gen 1 cycle. These included a change of lubrication to a higher temperature type, a new class of heat sink for an electronic circuit, addition of silicon grease under the heatsink to carry away heat and a change in circuit parameters to adjust the power switching point closer to zero, thereby lowering internal I^2R heating.

The initial Failure Rate **(Fig.8-7)**, Reliability Growth **(Fig.8-8)** and MTTF **(Fig.8-9)** forecasts are shown in the following graphs. Note that %FR numbers on the Failure Rate graph correspond to the value of DOTS on the Weibull graph **(Fig.8-6)**.

TEST 1				1 Γ =			PROJECT NAME?				
TOTAL	20			1.0		Acceleration < Factor	Line color 255, 0, 0	Beta Est_1 >>	1.80	1	83,790
Line #	Qty taken Off Test	Failure=1 Suspension=0	Rank Individual TIMES or CYCLES at Test Termination	Increment		Mean Order Number	Previous Mean Order Number	Median Rank %	Plot Point Y	Plot Point X
1			MILES			1.00				
1	1	1	20,720	1.0000		1.000		3.406%	-3.3622	2.07E+04
2	1	1	42,350	1.0000		2.000		8.311%	-2.4445	4.24E+04
3	1	1	55,860	1.0000		3.000		13.216%	-1.9537	5.59E+04
4	1	1	72,450	1.0000		4.000		18.120%	-1.6099	7.25E+04
5	1	1	83,790	1.0000		5.000		23.025%	-1.3406	8.38E+04
6	15								-1.3406	8.38E+04
7										

Fig. 8-5

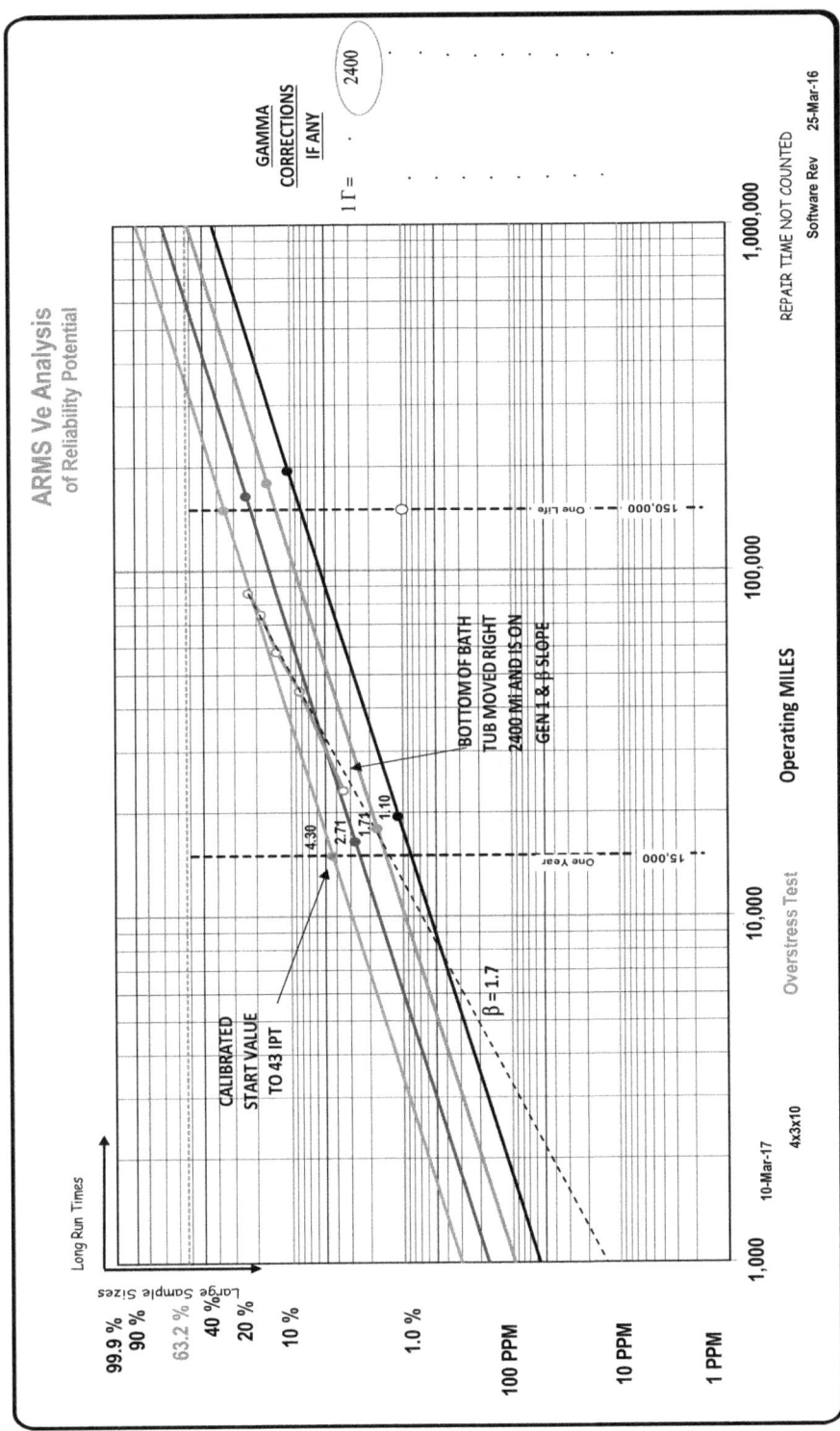

Fig. 8-6

163

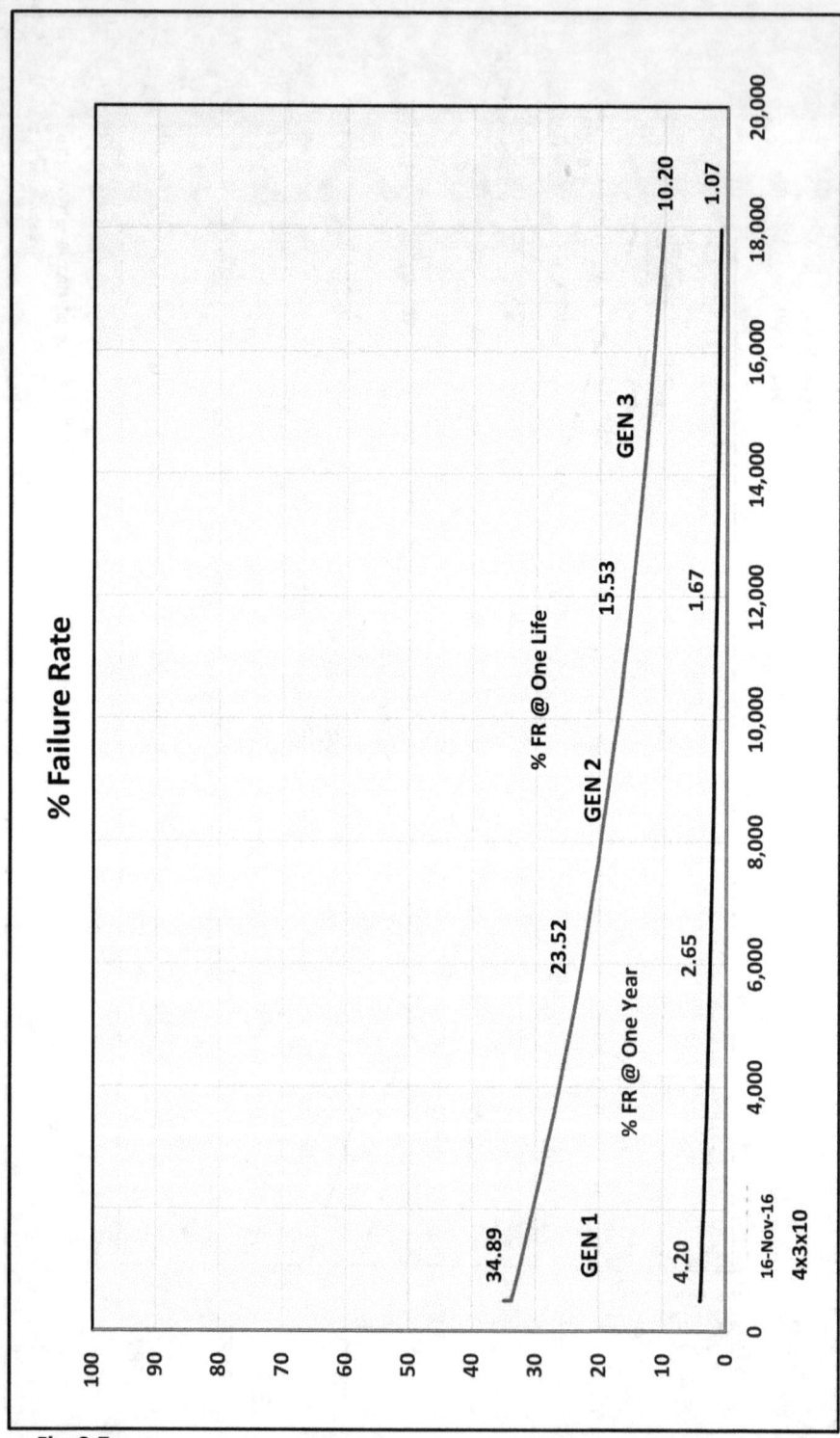

Fig. 8-7

164

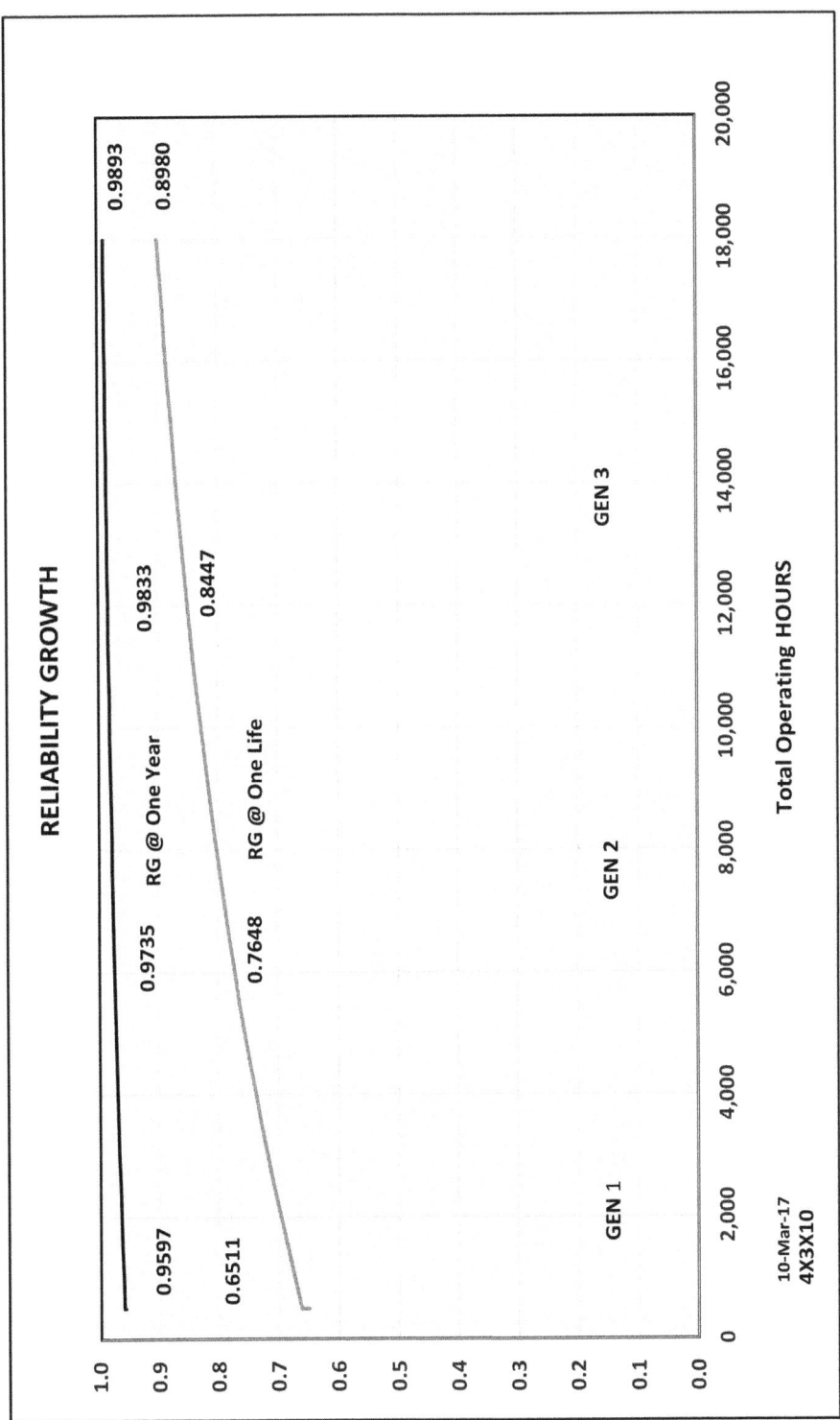

Fig. 8-8

165

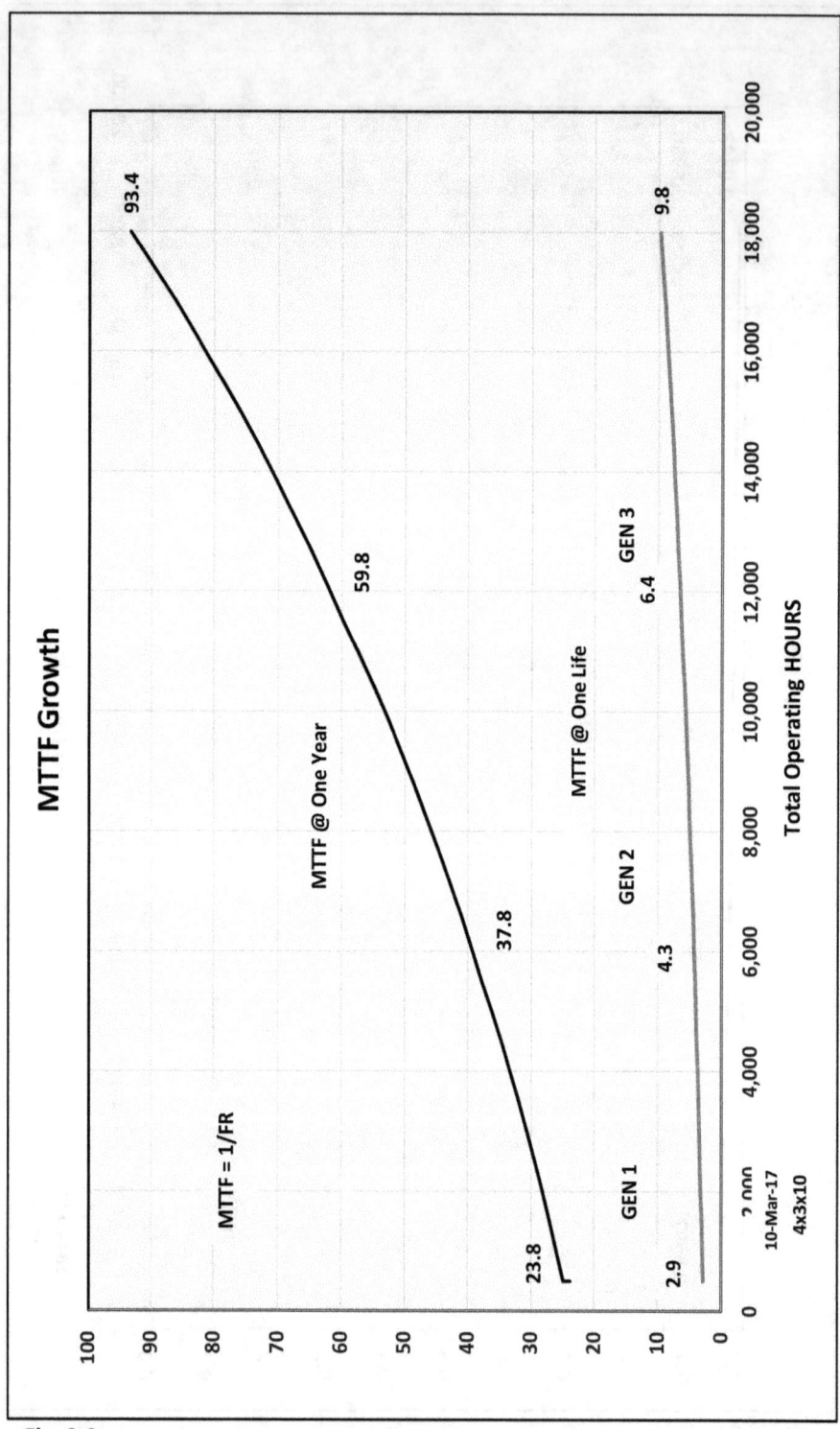

Fig. 8-9

166

In this case, the FR **(Fig. 8-7)** at:

One Year +extended hours: 4.7%, 2.65%, 1.67% and 1.07%

One Life +extended hours: 34.9%, 23.5%, 15.5% and 10.2%

The team believes these values are achievable if the "Assignable Actions" work is completed successfully and the durability testing is also completed as planned. With the ARMS' Weibull graph a % FR at other Times or Distances may be read directly from where they intersect the ARMS' Weibull graph forecast lines for Initial Design, GEN1, GEN2, & GEN3. Getting the hard work done can become a big "IF" issue. In this case, the durability test stand was run for 83,790 miles before ending the test, and 5/20 failures occurred as shown in the data entry table **(Fig. 8-5)**. This information was entered into the ARMS' Validation_Distance sheet and the ARMS' Weibull graph shown was obtained. The data plotted appears to be transitioning from an initial slope of less than one to a slope approaching one, however, we also know that there was about 80 hours of prior test time on the units. This prior time is accounted for by entering a "Gamma" value of Γ 2400 miles into the Validation_Distance entry sheet containing our test data. Gamma values entered will appear on the right edge of the Weibull graph when the ARMS' model is UPDATED. The gamma addition shifts all data to the right by 2400 miles. Because we are on a LOG scale, the wearout section appears a little steeper.

A vector was added to the ARMS' Weibull graph **(Fig. 8-6)** and adjusted to fit the slope of the wearout section by making entries to the Validation_Distance sheet. The vector is an estimate of the wearout test data beta slope and equals 1.7 in this case. The beta value estimate was arrived at by trial and error, entering values until a good visual match of the lines occurred. The vector can be

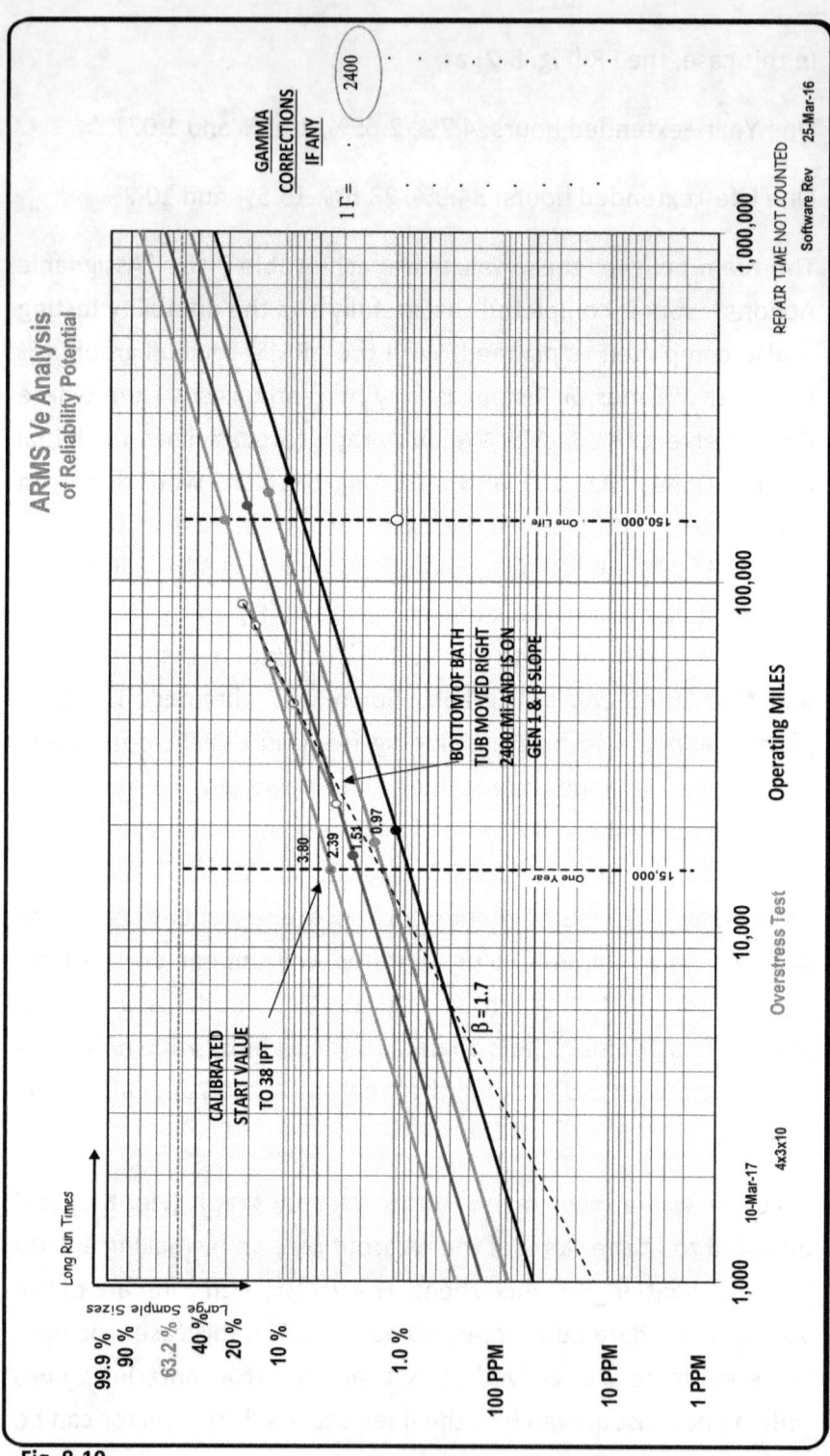

Fig. 8-10

168

used to match the characteristic wearout slope. If wearout has not been reached, the slope will be matched to the random failure slope during useful life. You will need to press UPDATE on the INPUT screen after a changed beta value to see a new result.

The shift in validation data due to the Gamma adjustment requires a "re-calibration" of the start value to a figure of 38 IPT. This adjustment returns the contact between the lowest value of validation data and the GEN 1 line **(Fig. 8-10).**

When all durability test data points are on the wearout section of the ARMS' Weibull graph, you do not know how deep the bath tub will be. It will be at least as deep as the lowest data point. The vector we added suggests a projection of the wearout line that would occur with a larger test sample size. This information can be used to estimate if you are getting closer to your goal, even though you have not tested the larger sample sizes. To actually demonstrate the values lower on the graph requires testing the costly larger sample sizes of a design you are in the process of changing. With this vector, a simple quick estimate can be made of failures to be expected at earlier operating times if a large population were tested or placed into service. In this example, the estimate would be approximately 89 PPM at 100 hours or 3000 miles of service. Obviously, not nearly good enough.

Chapter Nine

Entering Test Data

Into ARMS'

Weibull Analysis Graphs

- Dr. Fermi would agree that ARMS is about helping engineers to make better decisions and to get the necessary resources, not about making more precise projections.

- Facilitating Outstanding Leadership / Teamwork between all departments

- Engineering Design & Innovation Under Control

Data Entry Sheets

Data from durability tests or field applications may be entered into test data sheets and displayed on an ARMS' Weibull Graph. A Validation Time Tab and Validation Distance Tab support the two ARMS' Weibull Graph types. The Validation Distance Tab may instead be used to enter data in terms of cycles. A single ARMS' Weibull output may be expressed in one of three modes (time, distance or cycles). There are two separate Weibull graphs in the ARMS' model; %FR vs Time and %FR vs Distance or Cycles. Any units of Time or Distance can be defined. On the INPUT sheet, enter MILES or CYCLES into cell C6 to set the ARMS' model for the desired type.

There are ten data entry sheets each for the two graph types. (Weibull_Time and Weibull_Distance). This total of 20 data sheets may be viewed by zooming out and scrolling the screen. The upper row of sheets (#1-5) support up to 25 entries each and the lower row of sheets (#6-10) support up to 50 entries each. Failures that occur at the same time are grouped in an entry. Make use of the Excel comments function to enter any conditional information from the test program, field applications or failure analysis, e.g., overstress, extended testing, sample numbers, test stand utilized, customer data, gps locations, etc.

After test data has been entered into the first three columns of the data sheet, press UPDATE on the INPUT sheet to plot on the associated Weibull graph. Each line will be color coded to match the data sheet they came from. If data is entered into each of the ten data entry sheets for one of the Weibull types, there will be ten separate plots on one ARMS' Weibull graph. I recommend that you keep it simple and limit how many are shown at one time. Having multiple entry opportunities can be very useful when comparing different sets of test data.

171

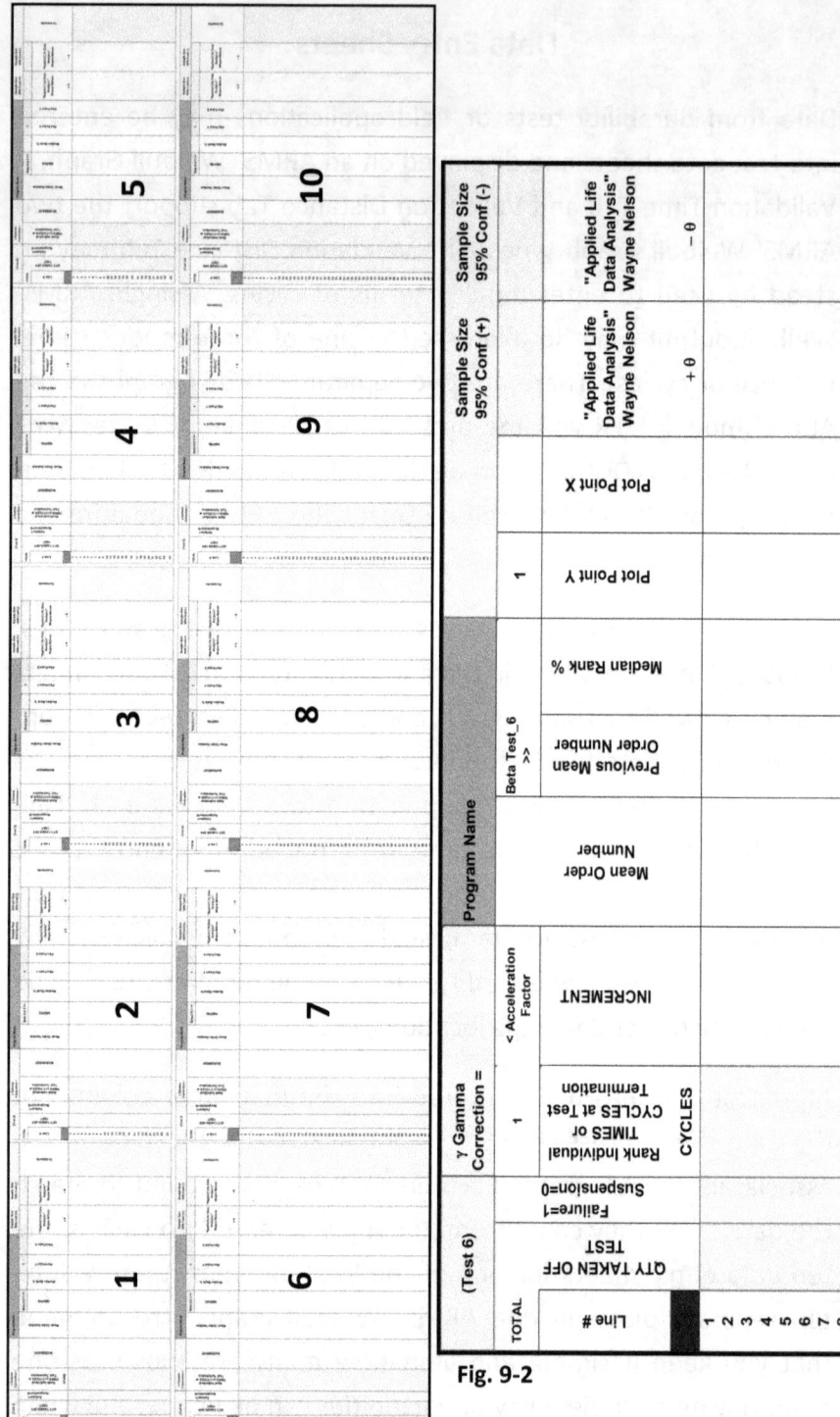

Fig.9-1

TOTAL	Line #	(Test 6)		γ Gamma Correction =		Program Name						Sample Size 95% Conf (+)	Sample Size 95% Conf (-)
		QTY TAKEN OFF TEST	Failure=1 Suspension=0	1	< Acceleration Factor	Mean Order Number	Beta Test_ 6 >>		1			"Applied Life Data Analysis" Wayne Nelson	"Applied Life Data Analysis" Wayne Nelson
				Rank Individual TIMES or CYCLES at Test Termination	INCREMENT		Previous Mean Order Number	Median Rank %	Plot Point Y	Plot Point X		+ θ	– θ
				CYCLES									
	1												
	2												
	3												
	4												
	5												
	6												
	7												

Fig. 9-2

In Validation sheet cells L3 & M3, you may enter 1 or 0 to generate an approximate +/- range of the 95th% median based on test sample size and +/- 1.645σ. The ranges are normally turned off by setting L3 & M3 to zero, simplifying the Weibull graph. Use the median "range" to aid in explaining to what extent increasing sample size will be beneficial, then I would suggest removing them to avoid an overly busy Weibull graph.

The limits may be applied as (-) lower limits or both (+/-) limits. For most applications, the (-) lower limits are of greater interest because they are more conservative. Upper (+) limits tend to be overly optimistic for general use.

The approximate median ranges are displayed as Weibull functions similar to the median. ARMS plots the (+/-) 1.645σ Range as exponential values. The result appears as straight lines parallel to the nominal line. This approach actually makes the

List failures before suspensions if times are equal — Program Name

Line #	QTY TAKEN OFF TEST (Test 1)	Failure=1 Suspension=0	Rank Individual TIMES or CYCLES at Test Termination — γ Gamma Correction = 1 (HOURS)	INCREMENT < Acceleration Factor	Mean Order Number	Previous Mean Order Number >> (Beta Test_1)	Median Rank % (4.00)	Plot Point Y (1)	Plot Point X (1,197)	Sample Size 95% Conf (+) "Applied Life Data Analysis" Wayne Nelson (1)	Sample Size 95% Conf (-) "Applied Life Data Analysis" Wayne Nelson (1)
	20 TOTAL									+θ	–θ
1	1	1	296	1.0000	1.00	19.00	3.406%	-3.3622	296	459	191
2	1	1	605	1.0000	2.00	18.00	8.311%	-2.4445	605	938	390
3	1	1	798	1.0000	3.00	17.00	13.216%	-1.9537	798	1,237	515
4	1	1	1,035	1.0000	4.00	16.00	18.120%	-1.6099	1,035	1,604	668
5	15		1,197	1.0000	5.00	15.00	23.025%	-1.3406	1,197	1,855	772
6								-1.3406	1,197	1,855	772
7											
8											
9											
10											
11											

Fig. 9-3

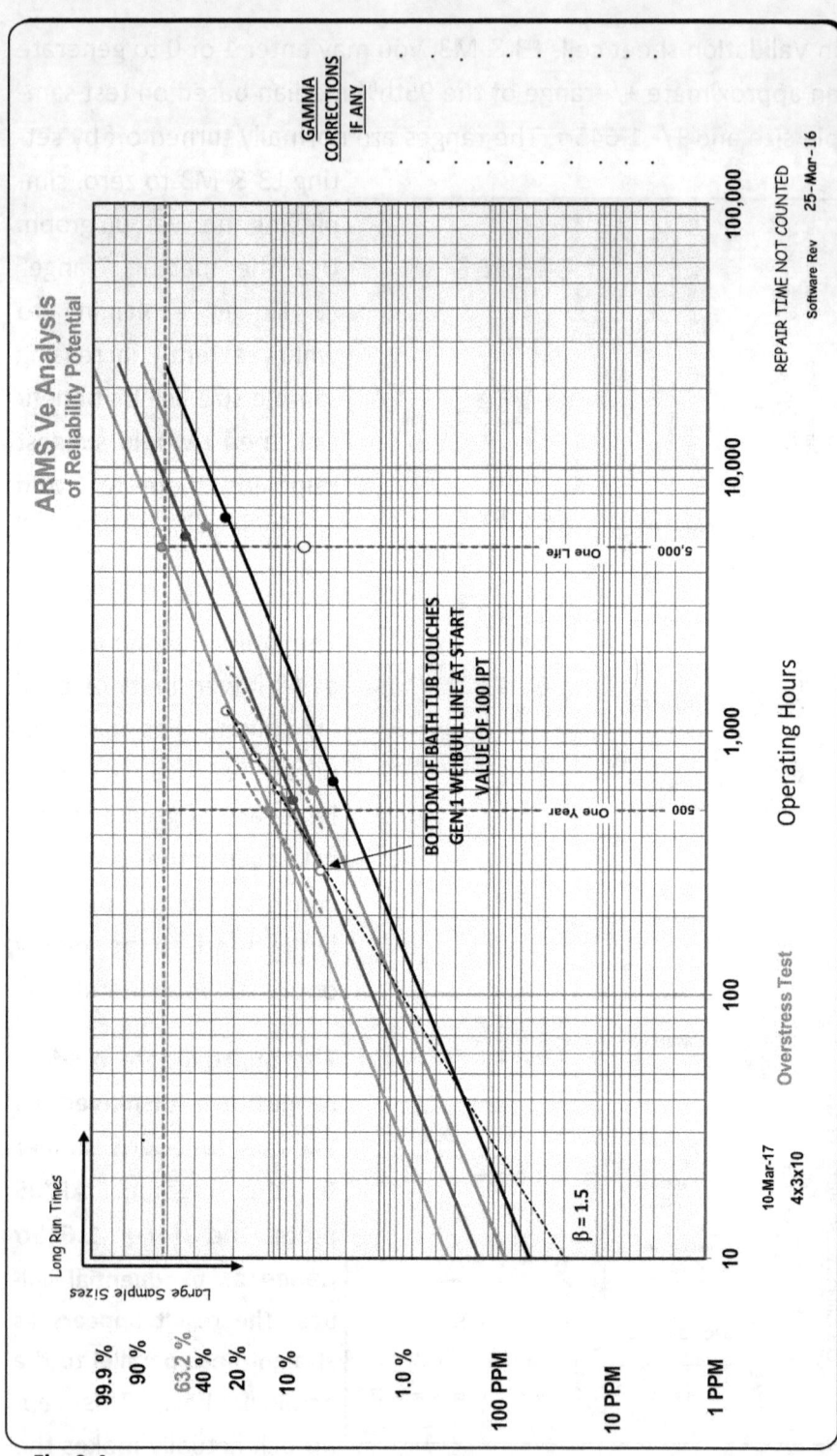

Fig. 9-4

range estimates useful when selecting sample size and anticipating the affect. However, project decisions should be based on the nominal median value rather than 95% ranges.

Dr. Fermi would agree that ARMS is about helping engineers to make better decisions and to get the necessary resources, not about making more precise projections.

Sample Size 95% Conf (-)	"Applied Life Data Analysis" Wayne Nelson	Sample Size 95% Conf (+)	"Applied Life Data Analysis" Wayne Nelson	Plot Point X	Plot Point Y	Median Rank %	NIBPSS	Mean Order Number	INCREMENT	Rank Individual TIMES or CYCLES at Test Termination	Suspension=0 Failure=1	QTY TAKEN OFF TEST	Line #
1	$-\theta$	1	$+\theta$	4,632	1		Beta Est.1 >>	1.00		1		27	2
	743		1,400	1,020	-3.6623	2.535%		1.000	1.0000	1,020	1	1	1
	1,785		3,362	2,450	-2.7512	6.186%		2.000	1.0000	2,450	1	1	2
	2,372		4,469	3,256	-2.2677	9.837%		3.000	1.0000	3,256	1	1	3
	2,769		5,215	3,800	-1.9318	13.488%		4.000	1.0000	3,800	1	1	4
	3,210		6,047	4,406	-1.6713	17.139%		5.000	1.0000	4,406	1	1	5
	3,286		6,190	4,510	-1.4564	20.790%		6.000	1.0000	4,510	1	1	6
	3,367		6,342	4,621	-1.2720	24.442%		7.000	1.0000	4,621		1	7
	3,375		6,357	4,632	-1.1093	28.093%		8.000		4,632		19	8
					-1.1093								9
													10

PROJECT NAME? *EXAMPLE 2 UNITS* *TOTAL* *Acceleration Factor* *Line color 0,0,0* *2Γ=1* *HOURS*

Fig. 9-5

Acceleration Factor: You may enter a positive only Acceleration Factor for this particular test. The application of an Acceleration Factor, however, is not a recommended practice.

The ability to do so was entered into ARMS at the request of Engineering teams as a backup to cover conditions learned after the ARMS' model was built. One factor is the team's performance. Teams new to ARMS and new to overstress setting may find an adjustment is needed across the model. This may be adjusted for many different reasons, but up to this point it has dealt with pessimism and optimism within the team dynamics. AF numbers will be in the form of 0.1 to 2 but should be set to 1.0 in nearly all but highly special cases (F ig.9-6 & 7).

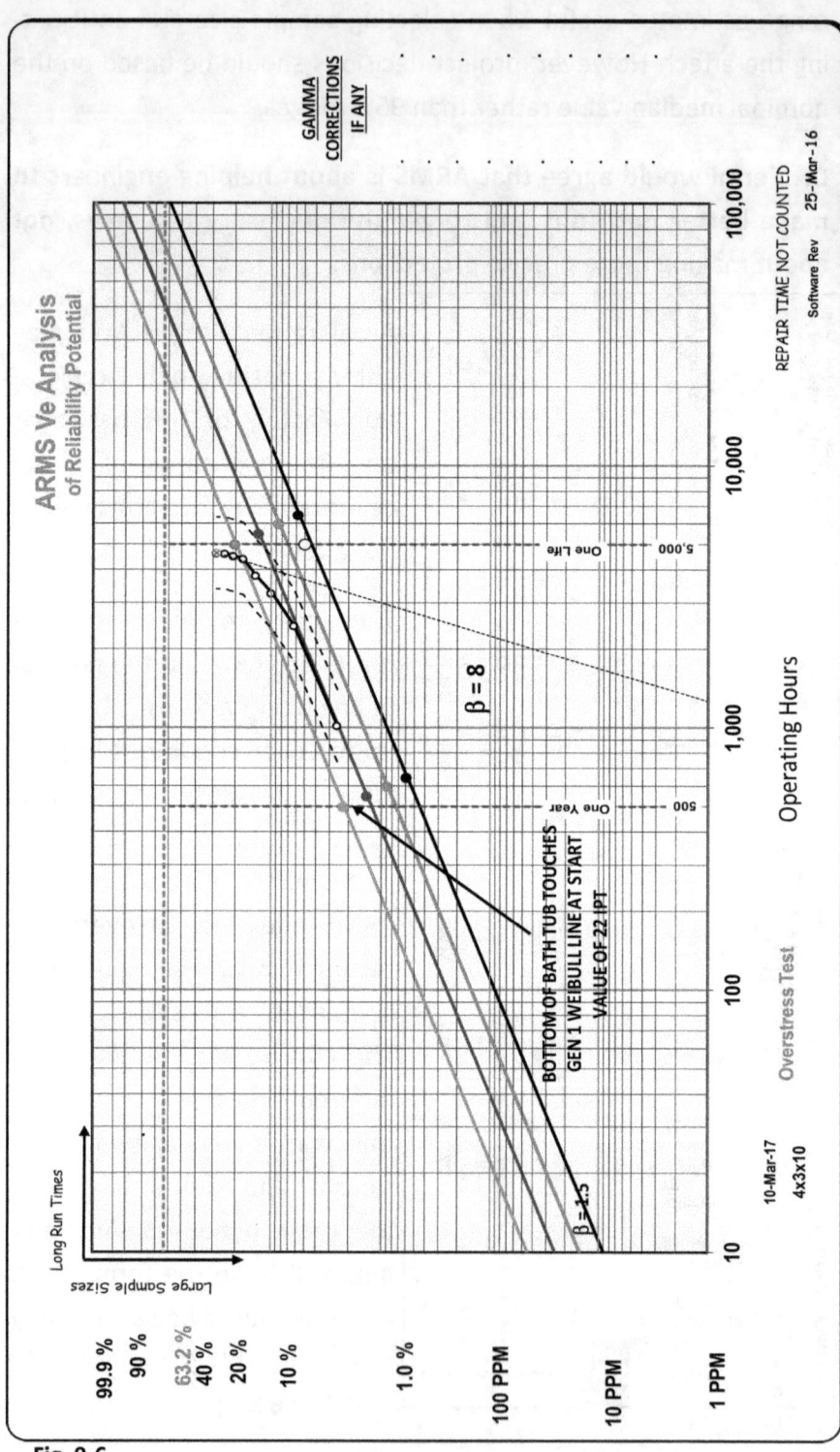

Fig. 9-6

176

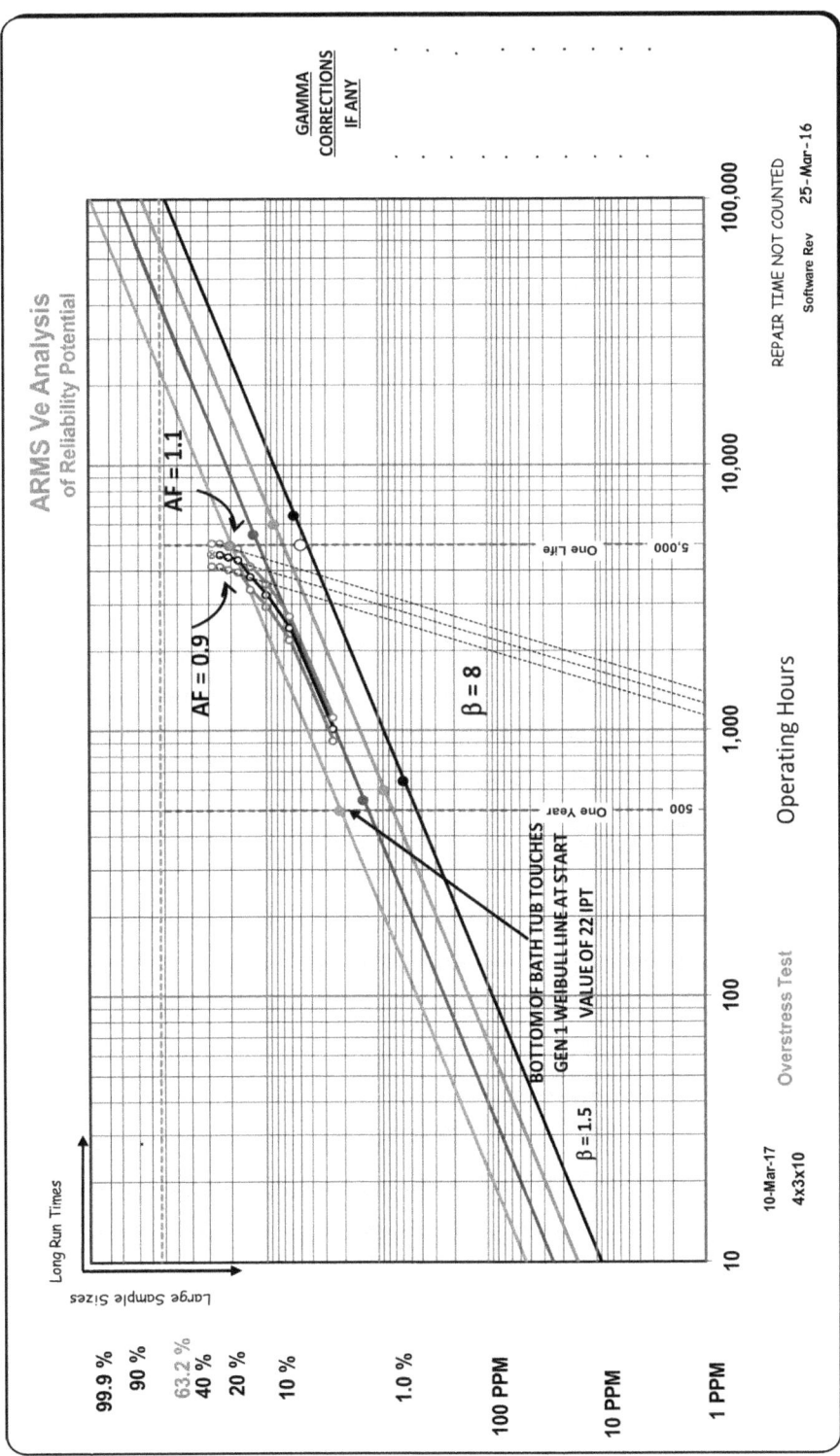

Fig. 9-7

177

Chapter Ten

Using Weibull Graphs with

ARMS

- Weibull graphs are unique tools that can improve our understanding of risk, reliability and validation test data.

- ARMS is a GUI interface between engineering visions, statistical tools, and measured results. ARMS helps engineers form, collect and document a vision and what steps are necessary to achieve the desired result.

- ARMS makes risk reduction processes visible.

- During advanced development and innovation, the collective wisdom of the team is far more important than any mathematical calculation we could possibly make.

- This ARMS' projection technique is how to get more information from your resource investments.

- There are no substitutes for knowing why something failed and what physics of failure was involved.

- Key questions for the Failure Analyst are why did the surviving units not fail?" and "Are there any signs of impending failure?"

WEIBULL GRAPH LOGIC.

TO UNDERSTAND HOW ARMS' WEIBULL PLOTTING RESPONDS TO CHANGEABLE INPUTS.

1) ONE YEAR AND ONE LIFE TIME LINES ARE AT THE HOURS, MILES OR CYCLES AS SHOWN ON THE INPUT SHEET.
   - NOT AFFECTED BY OVERSTRESS OR EXTENDED LIFE TESTING.

2) MAX HOURS AND MAX DISTANCE OF DURABILITY TEST PLANNED.
   - INCREASE WITH EXTENDED LIFE TESTING. (INDICATED BY LABELED DOTS AT THE END OF TEST.)

3) GOALS
   - ARE NOT EFFECTED BY OVERSTRESS OR EXTENDED TESTING.

4) ARMS' WEIBULL LINES
   - INITIAL CONDITION LINE IS NOT EFFECTED BY OVERSTRESS OR EXTENDED TESTS.
   - END OF GEN 1, 2 & 3 HAVE ENHANCED A LEARNING AND SPREAD WITH OVERSTRESS TESTING.
   - END OF GEN 1, 2 & 3 HAVE EXTENDED A LEARNING AND ARE SPREAD BY EXTENDED TESTING.
   _MAX TEST TIME AND DISTANCE ARE EXTENDED BY EXTENDED TESTS, IF COMPLETED.

5) TEST DATA PLOT POINTS
   - ARE ADVANCED BY ACCELERATED TESTS.
   - EXPECT MORE WEAROUT & FATIQUE TYPE FAILURES WITH EXTENDED TEST CONDITIONS.

6) - ACCELERATED TEST LABEL
   - ONLY SHOWN IF ACCELERATION IS GREATER THAN ONE.

OVERSTRESS CONDITIONS MAY HAVE MANY FORMS AND OCCUR SINGULARLY OR IN PLURAL.
'E.G. TEMPERTURE, VIBRATION, SHOCK, VELOCITY, ACCELERATION, HUMIDITY, CORROSION,
ANGLE OF IMPINGEMENT, VOLTAGE, CURRENT, FREQUENCY, ON/OFF CYCLE RATE, ETC.
WE WILL NEED TO ESTIMATE THE EFFECT ON EXPECTED FAILURES. 10% MORE(1.1) 30% MORE (1.3) ETC.
CAUTION: THIS APPROACH IS ONLY USEFUL WHILE FAILURES ARE DIRECTLY RELATED TO "AS FOUND IN NORMAL USE."

Weibull graphs are unique tools that can improve our understanding of risk, reliability and validation test data.

There are many excellent texts available through the internet to assist you in increasing your knowledge of Weibull applications. "Fatigue Testing and Analysis of Results" by W. Weibull is still available as are the writings of Dr. Abernathy.

ARMS provides the techniques required to assist engineers in making the right decisions on projects, capturing knowledge and to assist in getting the necessary resources. Weibull analysis is an important part of meeting this challenge.

As a visual representation of test results, the power of a Weibull graph will be enhanced if we know what to expect when durability data is entered. This can be illustrated using a Bath Tub curve. A cumulative Bath Tub curve graphed on Linear vs Linear, Log vs Linear and "log log vs log" graphs, will have the bath tub curve tipped up on the right side of the page **(Fig.1-8)**. Thus the Weibull lower left end of the tub is ascending at a slope of less than one ($\beta < 1$). The flat bottom of the tub on a linear graph has a constant slope near one ($\beta = 1$) and the far-right end, a slope up with ($\beta > 1$). In all stages, the cumulative failures are increasing. The ($\beta < 1$) area is known as the premature failure area and may contain processing defects. There may be several failures due to a design weakness causing a $\beta > 1$ at the start. I have seen this with designs that do not tolerate low supply voltage installations or other conditions that were not covered in the design requirements. The application and manufacturing anomalies are weeded out early.

Once premature failures are removed, the failure rate settles down to a near constant random rate with a slope close to one. This is referred to as the useful life period. Our objective will be to make the failure defined bath tub as deep as possible (a lower normal failure rate) and a bath tub as long as possible, i.e., longer useful life (Fig.1-10). At the end of life, the rate of failures and the Weibull slope increase rapidly with a (β>1) slope. β normally approaches a constant wearout rate characteristic of the product being tested. A Weibull graph is also useful when identifying bi-modal and multi-modal failure conditions. Weibull graphs visually allow us to fit the product's full life on one graph due to the compressed nature of the logarithmic scales.

ARMS is a GUI interface between engineering visions, statistical tools, and measured results. ARMS helps engineers form, collect and document a vision and what steps are necessary to achieve the desired result.

With ARMS, we will be graphing what the engineering team believes would happen if their design was built and placed into service. This engineering ARMS' assessment is based on a collection of inputs. In a process akin to Dr. Enrico Fermi thinking, the engineers mentally arrive at a rough estimate for a first-year failure rate. The Weibull graph of an ARMS' model forecast is shown as the constant useful life portion of a bath tub bottom having a slope of approximately (β = 1). The team does not yet know where the ends of their bath tub are located, nor do they know the slope of the wearout region. With some discrete components, there may be a recognized characteristic wearout beta slope. Ball bearings are a classic example of an item which fails with a characteristic wearout β for a given set of application conditions. Different bearing construction types; steel cage, full complement or non-metallic cage, have individual characteristic wearout slopes. When reported failures start to follow the characteristic failure rate, it is time to change all bearings yet running for similar times under similar conditions.

When samples are built and tested, they will be plotted on the same Weibull graph as our projection. Typically, the failures will be in the end of life phase and have a β slope greater than one. There may be some flyers early on that represent defects not consistent with the design and that make the Weibull failure distribution appear to be bi-modal. We will deal with bi-modals and data decomposition near the end of this chapter and throughout the ARMS' Handbook. There will generally be some random failures during the normal life period. These are interpreted after failure testing of the samples and as detailed failure analyses are being completed.

The Weibull graphical forecast shown (**Fig. 10-2**) indicates what the engineers expect from three generations of future design effort. This is based on an engineering team's "understanding" of the design, planned assignable actions, α Alpha rates of learning and **SF** Support Factors. There are four lines, (GEN 0) Initial Conditions (design as is), (GEN 1) after the first set of design improvements, (**GEN 2**) after a second round of improvement and (**GEN 3**) after a third round of improvement.

Each generation of design benefits from the results of the previous generations, build experience, test results, failure analysis and redesign implementations. Of course, if you follow a popular course of action by redesigning and building before test results are analyzed; projected gains may not be realized. The major improvement efficiency of ARMS results from repetitive cycles of team design reviews, formation of improvement plans and identifying & securing required resources. In this section, we will show how to work with Weibull graphs. Later, we will get into how to generate the adjustments shown here by using an ARMS' model for inputs.

The (GEN 0) initial **conditions line is only an educated guess, and will almost always be incorrect when compared to actual measurement of samples. However, it is not a given as to which is the**

182

most accurate. Test programs and executions are also subject to error. For example, one error may be that test results are on last week's or last month's design configuration and the engineer's new vision is of next week's or next month's design.

The important point is that the initial start failure rate is what the team believed to be the case based on the existing design and their thinking at the time it was chosen. In the case of an initial ARMS' effort, the team's assessment is the most accurate estimate we have and a vision the engineering team is working to improve. We do not want engineers working from a list. Rather, we are using the activity of forming a listing to help modify what the engineers are thinking as they work with their visions. ARMS documents visions in a recoverable file. The truth will become clearer during the following ARMS' discussions.

ARMS makes risk reduction processes visible.

We use the ARMS' process key functions; (α & **SFs)** to estimate the spacing between the bottoms of bath tub curves for an initial design and three following generations of design improvement. When validation test failure results are obtained, in most cases, we will be graphing the wearout portion of the design's bathtub curve. An ARMS' Weibull graph will bring the estimated and measured elements together by making adjustments to the engineering estimates in line with the measured realities. This feedback of knowledge will enhance engineering perceptions for the future. To a degree, it is like making a weather forecast. The weather forecast will help us decide on whether we should go to the beach tomorrow, but don't look back and try to link the forecasts together. Over time, a team's ability to forecast will improve due to experience with the feedback provided by ARMS.

As testing occurs, resultant failure data is plotted on the same

Weibull graph and utilized to "calibrate" the initial engineering estimate of the START failure rate by bringing the two into agreement at the lowest point of the measured data (**Fig. 10-3**).

Engineering teams are more accurate at estimating relative relationship between design generations than setting an absolute value for the **START** rate of failure. The use of relative valuations is a very important concept **(Fig.1-11)**.

During advanced development and innovation, the collective wisdom of the team is far more important than any mathematical calculation we could possibly make.

Mathematical calculations are a guide for the engineers, but should not overly restrict their ability to form new visions. Teams which struggle with innovation are teams whose visions are being restricted or not supported. Your visions need to be supported by science or you need to open new doors for science to grow. Remember when we had nine planets and life before plastic?

Because the engineering team is thinking 24/7, before we can complete and evaluate (GEN 1) by validation testing, the team will have already developed, and in some cases started to implement, a list of design changes for the next generation build. The same is true of our Weibull analysis calculations. We need validation test results and calculations to bring in a degree of reality. But as an engineer, I know we must always stay close to the engineering team because they are way ahead of us. Remember, the engineers are always in control of the design and the next configuration will come through them. If the engineers get too far ahead of validation testing and analysis, there could be a lot of wasted program effort and a need for additional design turns. It is a leadership judgement as to when to push for the next design cycle. ARMS can help in making these critical decisions.

When working with Weibull graphs, there are a number of data interpretations and solution paths that can be perceived. ARMS' analysis is no exception. A few concepts in our pursuit of a design solution follow;

1) ARMS is utilized by the engineering team to review the proposed design concept in detail. ARMS does this by breaking the design vision down into indented elements and sub-elements, assigning each line item a relative risk percentage, α and **SF** where needed.

2) **Ideas and new solution paths revealed during discussions are captured as assignable actions and Pareto ranked by relative risk, to form a risk improvement plan. This plan is focused on required steps to actually do something about solving identified risk issues. This approach to solving issues is the primary and most important output of ARMS.**

3) Perceived progress and expectations are analyzed with Weibull graphic techniques to aide in providing a new vision of the future. These visions reflect the thinking and engineering expectations at the moment they were developed. ARMS' improvement plans formed become a record of design progress. In the end, as forecasts, Weibull projections are almost never quite "right" by the time we obtain measurements, but the forecasts were perceived to be right when they were formed. The difference is corrected when we make calibration adjustments. The "Assignable Actions" developed lead the team in the right direction to lower risks.

4) When validation test data becomes available, it is used to adjust the START value position the (GEN 2) cycle forecast next to the knee or lower end of the Weibull plot of measured data. The data plot is a bath tub section and the knee or lowest

measured value is where wearout begins in the sample. This plotted position represents the "demonstrated value of reliability risk."

5) ARMS supports adding a vector to the wearout portion of the validation test data. The vector starts at the highest measured value and is manually aligned to the wearout section of data forming the beta slope for the characteristic wearout phase. The lower projection of this vector indicates what should be expected if larger samples of this design generation were exposed to test conditions. At "One Life" or any other point along the horizontal axis, the depth of a failure demonstrated bath tub depends on the sample size involved.

To demonstrate and plot a 1% failure rate requires at least 1 failure out of 100 tested. There are also Bayesian techniques, not covered here, that can help with the limitations of small sample sizes. (See Amazon, Bayesian Data Analysis, 3rd edition, Gelman, Carlin et al.)

With data from measured physical items, we create a "demonstrated bath tub curve." By applying a vector to the wearout data from testing physical items, we create the right hand end of an "estimated deeper bath tub" similar to what you would obtain using a larger test sample. **With obvious reservations, it is a way of projecting an estimated result without incurring the additional expense. This ARMS' projection technique is how to get more information from your resource investments.** Use it with caution.

When following an ARMS' Weibull approach, there must be support, from an "excellent failure analysis team," analyzing validation program failures and survivors. Think of the NCIS Television program! You need some "Abbeys" on your forensic team. The benefits derived from the detailed analysis of validation test survivors cannot be over-emphasized. Knowing why items fail is half of the question; you also need to know the other half, why others survived. Spending money in this area is a value added expense.

Let's look at some generic validation test data, the results from 5 failures. The test results are entered into a "**Verification_Time**" data sheet and are plotted on a Weibull graph when the ARMS' model is **UPDATED** on the **INPUT** sheet. The rough initial forecast from the associated ARMS' Workshop did not agree with the measured values obtained from validation testing **(Fig.10-2)**. This is normal at this point in the ARMS' process. It is necessary to "calibrate" the ARMS' Workshop estimates. As mentioned earlier, calibration is accomplished by adjusting the initial "**START**" value estimate on the "**INPUT**" sheet, up or down as needed. With **START** value adjustments, the (GEN 1) Weibull line is shifted until the line just touches the lower end of the measured data creating a "knee" in the curve **(Fig.10-3)**. This is where our demonstrated "Bath Tub" starts to slope up and β is > 1.

A "characteristic wearout vector" was applied and adjusted to a slope of 2.5 matching the measured wearout section.

Increasing the initial "Start Value" estimate to 165/1000, the (GEN 1) line, with a slope approaching one touches the bottom end of the test data. (See arrow on **Fig.10-3**) The (GEN 1) line now also crosses the one year dashed vertical line at about 9.5% failure. The (GEN 2) and (**GEN 3**) lines cross one year at 6.0% and 3.2% respectively. The (GEN 1) line is now "calibrated" to the validation test data. The end of test for GEN 1-3 is represented by the DOTS on the ARMS' Weibull graph, and are 16.5, 10.7, 6.97, and 4.2 **(Fig.10-3)**.

The vector with a slope of 2.5 crosses the one-year dashed line at about 2.8% FR. This is interpreted as "if a larger sample of the

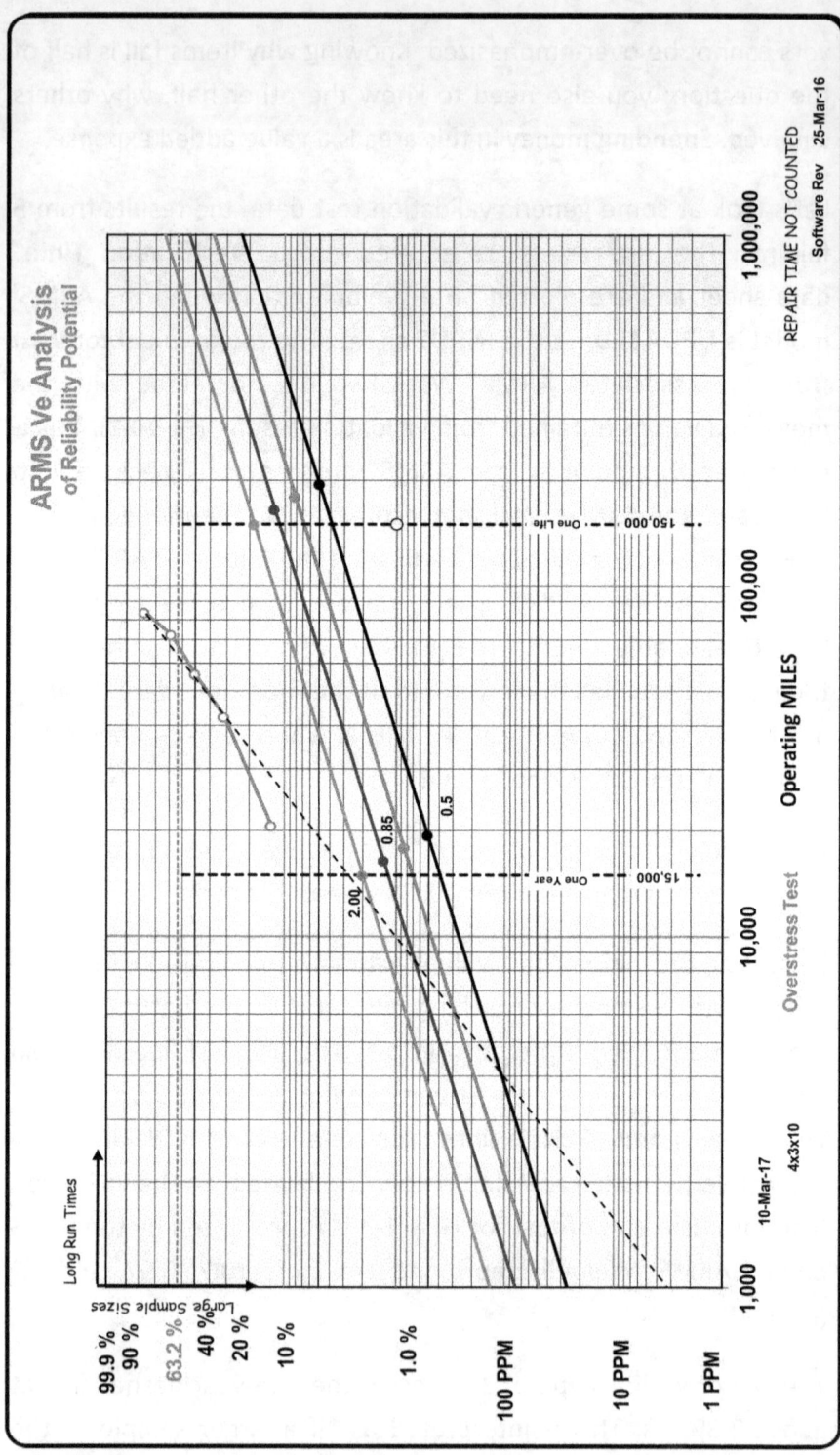

Fig. 10-2

188

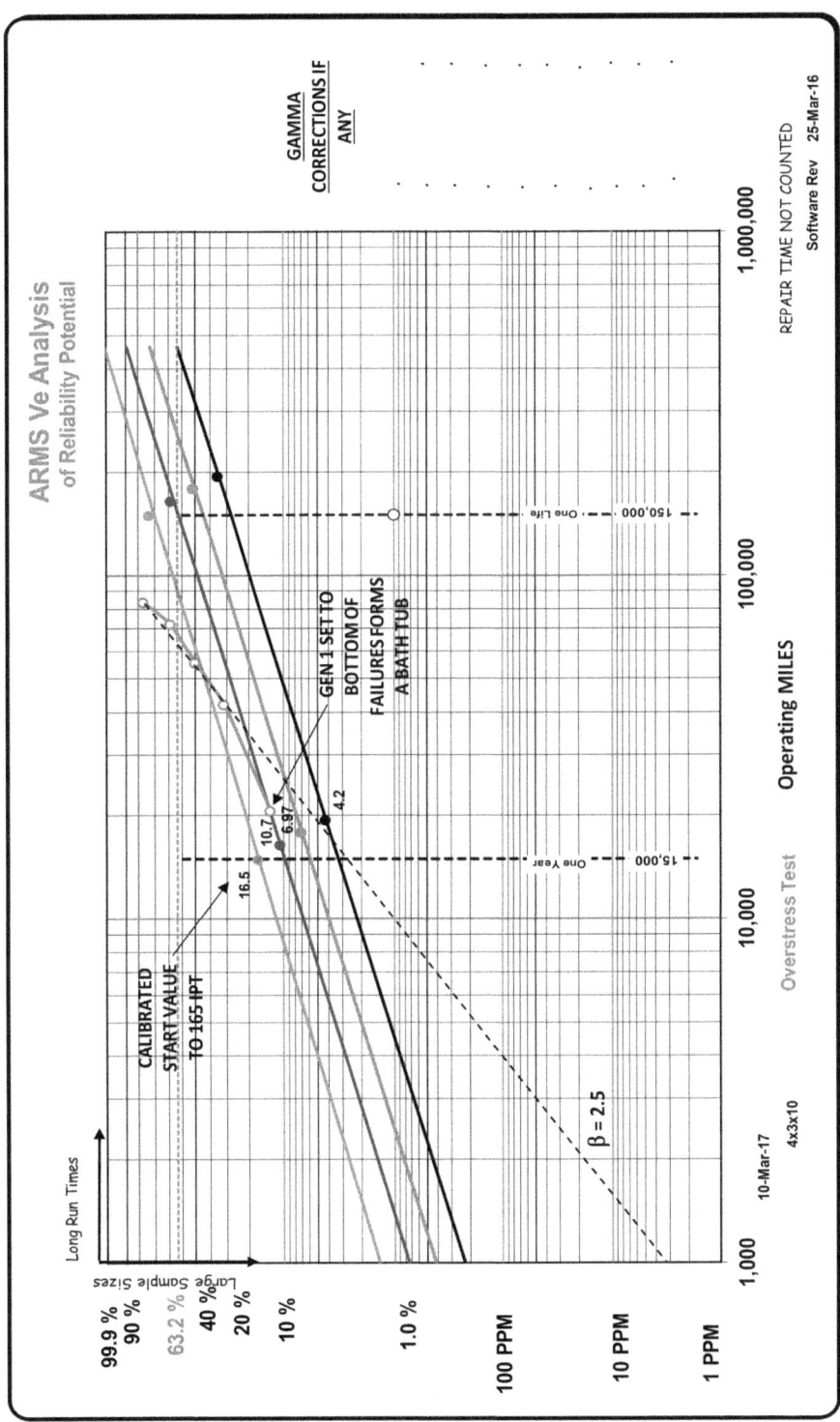

Fig. 10-3

189

same design had been tested, it would have a median failure rate at one year of 2.8 %." Remember, these are rough adjustments to make engineering level estimates, so do not get carried away by the process. Keep focused on solving the "Assignable Actions" and "Support Factor" problems.

At this point, it is difficult to say very much about our ability to achieve the (5% @ 150,000 Miles) Advanced Development Goal. This ability will improve as we proceed. It is clear that we need to advance the characteristic wearout line to the right until the vector is to the right of the goal. As we proceed through additional design cycles, our sample sizes normally increase and our estimates will improve, making the rate of progress clear. Estimates will come from analyzing validation test units, completing corrective "Assignable Actions," jointly reviewing all that has been learned and benefiting from special problem-solving techniques applied to identified failure issues.

By now, some ugly little revelations about working with any Weibull graph may occur; sample sizes and long run times necessary to demonstrate that the goal has been met may not be physically or financially possible. Small sample sizes and short run times are both **SF** issues. If you end up stuck in this box, seek extended tests on samples that have survived and use combined multiple overstress environments. If you sensibly fail units faster, you can get in more design turns with the same budget.

There are no substitutes for knowing why something failed and what physics of failure was involved.

Work on the **SF** issues, but also investigate "outside" help, scientific reviews, Design of Experiments, Shainin Red-X, Goldfire Innovation, Finite Element Analysis, TRIZ and specialized bench or token tests which cost less and can be conducted in volume. Often you cannot overstress test beyond an identifiable point due to creating failure modes unrelated to reality. In those cases, you may be able to isolate multiple overstress testing to sub-systems or sub-sections of the design. ARMS' discussions within the team on these issues and actions is invaluable. Often there are multiple failure modes at work and Weibull analysis can reveal the condition and aid in identifying which groups can be analyzed in more detail for specific modes of failure.

The spread between ARMS' Weibull estimate lines is due to the α learning rates and applied (**SF**) Support Factors. In this generic example, our interim Advanced Development Goal is set at (5%) following a normal useful life of 150,000 Miles. This exit goal is the required level of development before turning the design over to Production Engineering for additional development. As previously stated, it is normal to average 12 design turns before the start of production. If we increase the velocity of design iterations, we may have more smaller turns in less time.

A detailed analysis of failures includes a review of all historical data leading up to durability testing, test conditions, symptoms of failure and any other conditions suggesting an impending failure. This failure analysis revealed that new test stands were also being developed at the same time as our product design and units placed on test were utilized in setting up, calibrating and conducting statistical process studies to qualify the validation test stands. During this period, the units were subjected to approximately 60 hours of operation, some of which were beyond specified limits. Regrettable, but things like this do happen. Failure analysis must look into all possibilities. The effect on the Weibull plot of test data is to cause a slight curvature in the lower tail end of the data **(Fig.10-3)**. The curvature, which does not follow the characteristic wearout dashed line, may be the result of undocumented actions by suppliers, the test team, possible burn-in activity, effects of poor fits and finishes, prolonged performance testing and other actions. Not all samples may have been at the same starting point when durability testing began. It is normal to expect the wearout failure data to be in a nearly straight line with a slope greater than one. To determine if distortions in the test data are occurring, it is necessary to have an intimate knowledge of the build, test and failure analysis activities. Failure analysis trips to suppliers by a qualified Failure Analyst are always productive.

The curvature was determined to be an indication of a need for a "Gamma" adjustment. A Gamma adjustment adds to the reported test period until the plot approaches a reasonably straight line and is still consistent with all analysis findings. In this case, there were approximately 10,000 miles of operation while setting up and qualifying test stands and while documenting initial performance values. Gamma adjustments are entered on the associated validation data entry sheet. After entering a Gamma adjustment and pressing "UPDATE" on the "INPUT" sheet, the ARMS' Weibull projection will need to be recalibrated to align the measured and forecasted information. Adjustments to the "START" value will be required. In this case study, "START" becomes 110 IPT which is 11%. When a Gamma adjustment has been applied, it will be displayed on the right edge of the ARMS' Weibull graph (**Fig. 10-4**).

Use Gamma adjustments with caution, because you may not be dealing with a normal curve in the "Bath Tub," or there may be bi-modal or multiple failure causing conditions in the test samples or test runs which require decomposition first. When multiple builds and/or test stands are in use, they need to be documented, controlled and be a part of the failure analysis.

On the ARMS' Weibull graph (**Fig. 10-4**), the top Initial condition

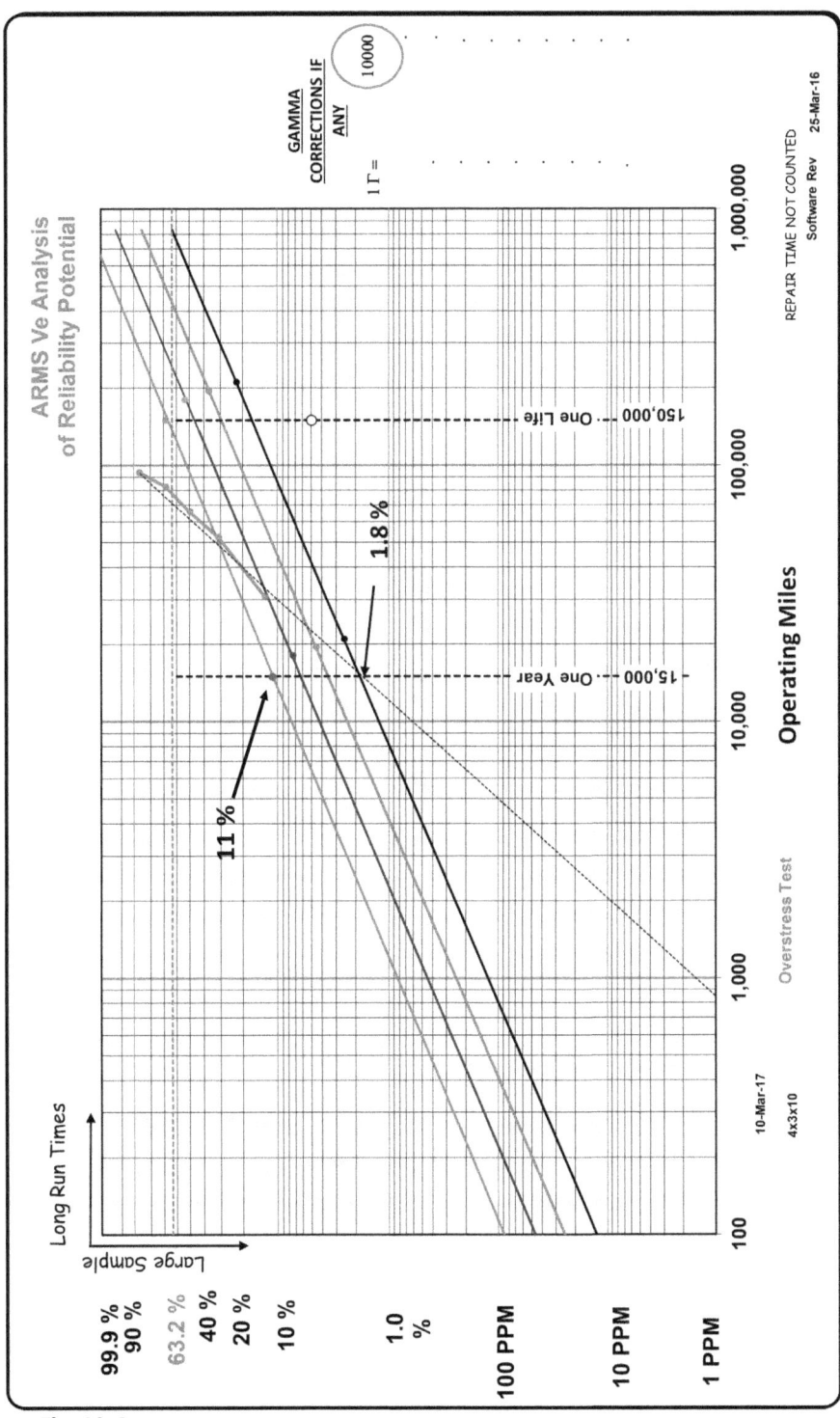

Fig. 10-4

193

line indicates 11% where it crosses the first-year line, (GEN 1) crosses at 6.3 %, (GEN 2) at 3.5 % and (GEN 3) at 1.8%. By pure coincidence, the wearout vector projection from the measured data also crosses year one at 1.8 %.

As work proceeds, projections made by the engineering team become more relevant. With more design iterations, the accuracy of engineering projections continues to improve. The more ARMS' projects engineers are involved with, the more their ARMS and engineering performance improves. ARMS is a learning process for the team around the table and ARMS also peaks individual engineering abilities. The search for greater opportunities to im-

TEST 2			$2\Gamma =$		PROJECT NAME?					
TOTAL	9		1.0	Acceleration < Factor	Line color 0, 0, 0	Beta Est_2 >>		7.00	1	90,000
Line #	Qty taken Off Test	Failure=1 Suspension=0	Rank Individual TIMES or CYCLES at Test Termination	Increment	Mean Order Number	Previous Mean Order Number	Median Rank %	Plot Point Y	Plot Point X
2			MILES		1.00				
1	1	1	6.80E+04	1.0000	1.000		7.413%	-2.5637	6.80E+04
2	1	1	7.50E+04	1.0000	2.000		18.059%	-1.6136	7.50E+04
3	1	1	8.10E+04	1.0000	3.000		28.706%	-1.0836	8.10E+04
4	1	1	8.80E+04	1.0000	4.000		39.353%	-0.6929	8.80E+04
5	1	1	9.00E+04	1.0000	5.000		50.000%	-0.3665	9.00E+04
6	4							-0.3665	9.00E+04

Fig. 10-5

prove projections usually leads to new levels of thinking and breakthroughs. Learning and design progress is not a continuum, rather a series of steps, jumps and leaps.

Another set of 9 with 5 failures, (Fig. 10-5) and (Fig. 10-6), had failures that were in a state of wearout. There were no Gamma adjustments and the START value was "calibrated" to 29 IPT as shown. Measured wearout data was set to just touch GEN 1 line. This is a straight forward analysis of a normal single mode failure pattern.

Real life test samples often contain multiple failure modes. Small

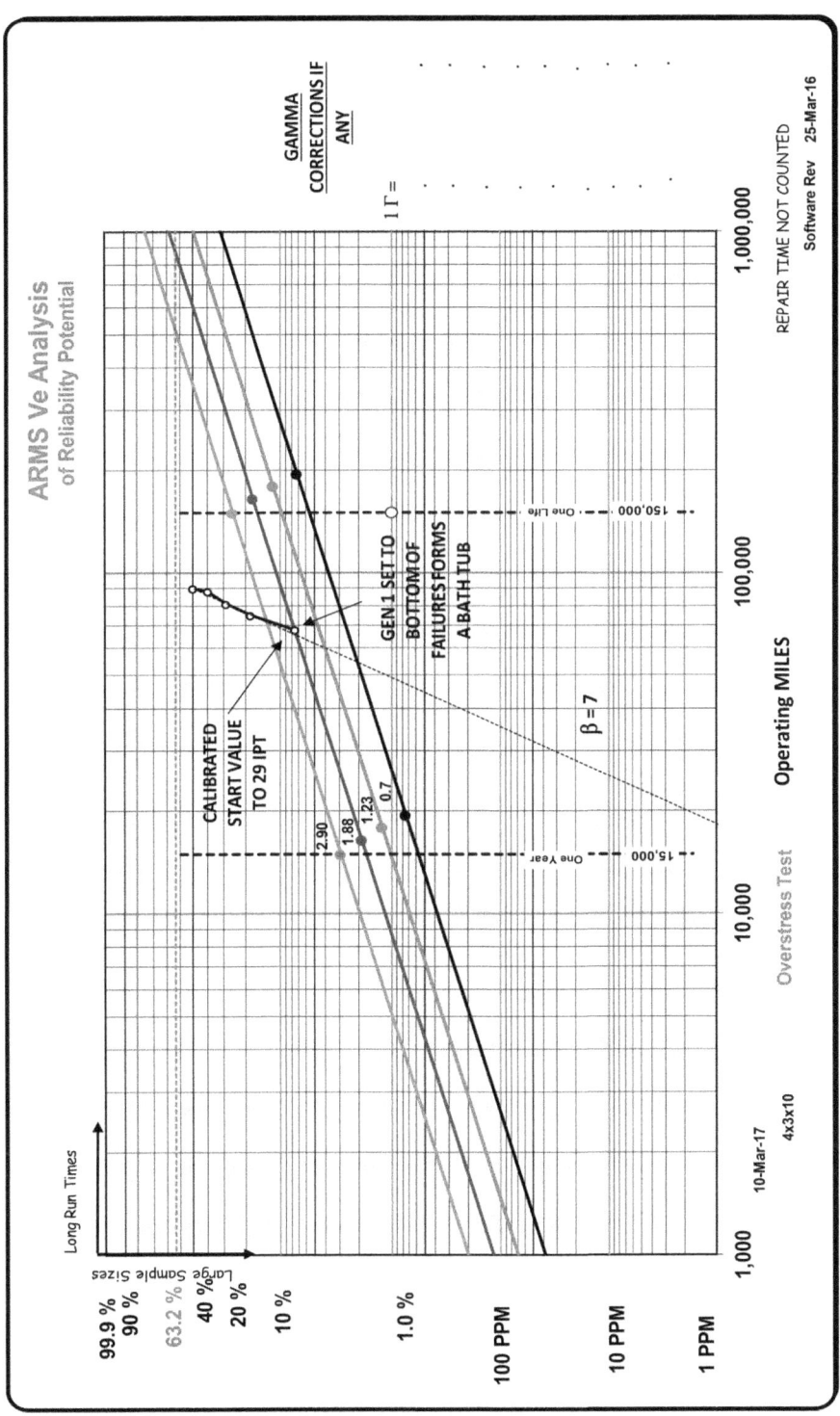

Fig. 10-6

195

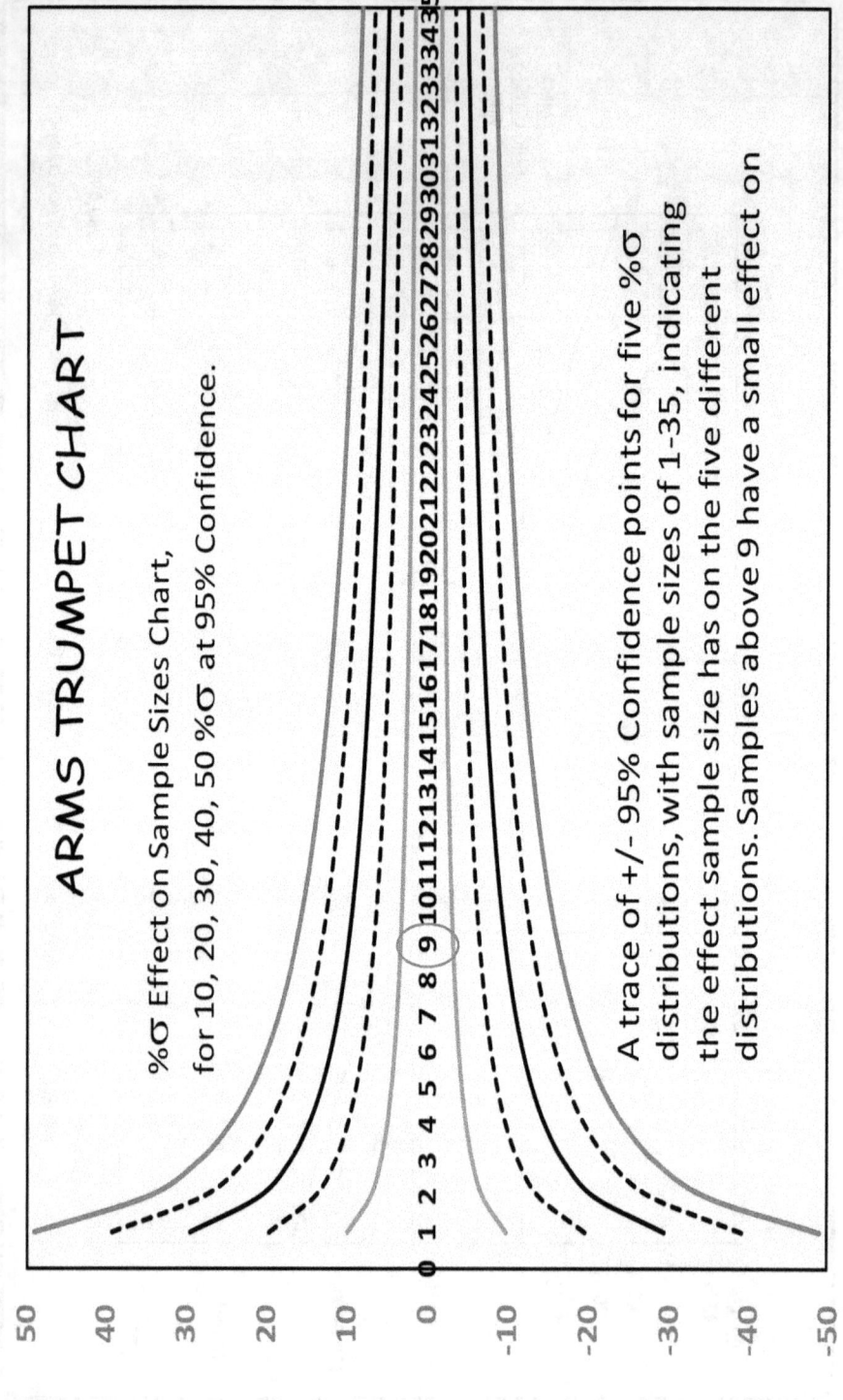

Fig. 10-7

196

validation test sample sizes, by chance, may not contain one of the relevant failure modes when multiple modes are present. The small sample will not support valuable post test data analysis, utilized to direct failure analysis efforts, and lead to insight and a vision of required solutions. The tradeoffs involved in test sample sizes are many faceted. As a general rule of thumb, **I agree with John Dukro in preferring a sample size of nine units for most validation work.** If it is known that multiple failure modes are present, then a larger sample size than nine will be required to support decomposition following testing.

Review of the Trumpet Chart **(Fig.10-7)**, indicates that beyond a sample of nine we enter into a region of diminishing returns. Some design development programs call for larger sample sizes that are required to capture random fault conditions and to evaluate multiple failure modes. The Lithium Battery (LBS) and Fuel Injectors, which require very low failure rates, are examples of designs requiring larger sample sizes.

When a learning path appears to behave like a slow continuum and the more we work on it, the harder it is to realize improvement, our true Alpha learning rates have reduced with each new cycle. If you detect that this is happening, seek "out of the box" thinking, breakthroughs, brainstorming or consider design of experiments applications, Goldfire Innovator, TRIZ, and It may also be time to consider the "Queen of Hearts" solution, if you can't solve the problem, change the design to one you can.

When special tools are to be applied, color code the ARMS' model Input sheet cells to reflect what, when and where and these actions are to be taken and allocate resources for these extracurricular activities.

Design generations, as utilized in ARMS' discussions, are major iterations of the design, usually in line with budgetary cycles. Within a design generation, there normally are several adjustments to the design based on knowledge gained during the build and early tests.

We need to evoke projections, overstress tests and extended life tests to the extent possible. Testing for more than one life time can extend the usefulness of samples, provided they survive long enough and fail with modes of failure consistent with normal life. Detailed failure analysis must confirm conditions are consistent with normal life. You need to also consider conducting overstress tests which are intended to force early failures, provided the failures are still relevant to real applications. Forced overstress failure testing can be very insightful and speed up the validation process with fewer samples.

ARMS has not attempted to define the parameters of overstress tests due to the many case specific variables that can only be assessed for each program by the program Validation Engineers. ARMS does flag the fact that overstress tests are being conducted, when true, to alert the reader that overstress conditions are being utilized. The effect of overstress testing is applied by the Validation Engineer on the Input Sheet. A program Validation Engineer needs to participate in and contribute to the ARMS' Workshop for a relevant period.

There are two vertical dashed lines on an ARMS' Weibull graph, One Year and One Life. These result from the input data. Where One Year crosses the projected Weibull lines are estimates or forecasts of the first-year failure rates. While not yet good enough for production, during advanced development we need to search for missed opportunities and design breakthroughs that will be required. With ARMS, the objective is to confirm and improve knowledge of design deficiencies earlier in the budget cycle. ARMS' information can trigger special actions on high risk sub-components. Another strong use of an ARMS' Workshop is to have the team compare two or more design concepts or two supplier's concepts with separate ARMS' models. An ARMS' Weibull approach has been utilized when down selecting competitive sources. Risks as well as initial cost are evaluated.

As a product enters into the wearout phase of its life, it may exhibit a β slope characteristic to the product, e.g., ball bearings often exhibit a wear out β slope of 1.5. If historical data exists to support a product characteristic slope of 1.3, 1.7, etc., we may evaluate expectations lower on the graph than revealed by the forecast. With ARMS, we can project the slope of the failures and determine their β angle.

An ARMS' Weibull graph (Fig. 10-9) involving the five failures and the four suspensions 5/9 follows. The Characteristic wearout vector added to the final three failures has a β slope of 4.5. The addition of the vector clarifies the condition that early failures are of a different nature than the final cluster of failures. The "inverted S" shape is a strong indicator of multiple failure modes. Decomposition of the data into a set of 2 and a set of 3 and plotting the results clearly reveals separate failure modes. Two tables of data are shown, (Fig. 10-10) for the 2/9 failures with a β slope of 2.2, and a second table (Fig. 10-11) for the 3/7 failures with a β slope of 7. Attention is focused on the 3/7 failures which are in a wearout condition. The 2/9 are treated as premature failures and will be analyzed independently. The ARMS' model is calibrated by setting GEN 1 to the lower end of the 3/7 wearout line with a START value of 35 IPT (Fig.10-12).

The problem with this analysis is that the samples are small, as previously stated. Anticipating issues of this type needs to be entered into the thought process when selecting sample sizes.

(Fig. 10-8, 9, 10, 11, 12) provide a clear indication that the wearout end of a bathtub has been reached. Analysis of surviving units for

PROJECT NAME?

	TEST 2		$1 \Gamma =$	Acceleration		Line color 0, 0, 0	Beta Est_2 >>	4.50	1	88,000
TOTAL	9		1.0	< Factor						
Line #	Qty taken Off Test	Failure=1 Suspension=0	Rank Individual TIMES or CYCLES at Test Termination	Increment	Mean Order Number	Previous Mean Order Number	Median Rank %	Plot Point Y	Plot Point X	
			MILES		1.00					
2										
1	1	1	1.95E+04	1.0000	1.000		7.413%	-2.5637	1.95E+04	
2	1	1	3.01E+04	1.0000	2.000		18.059%	-1.6136	3.01E+04	
3	1	1	7.20E+04	1.0000	3.000		28.706%	-1.0836	7.20E+04	
4	1	1	8.00E+04	1.0000	4.000		39.353%	-0.6929	8.00E+04	
5	1	1	8.80E+04	1.0000	5.000		50.000%	-0.3665	8.80E+04	
6	4							-0.3665	8.80E+04	
7										

Fig.10-8

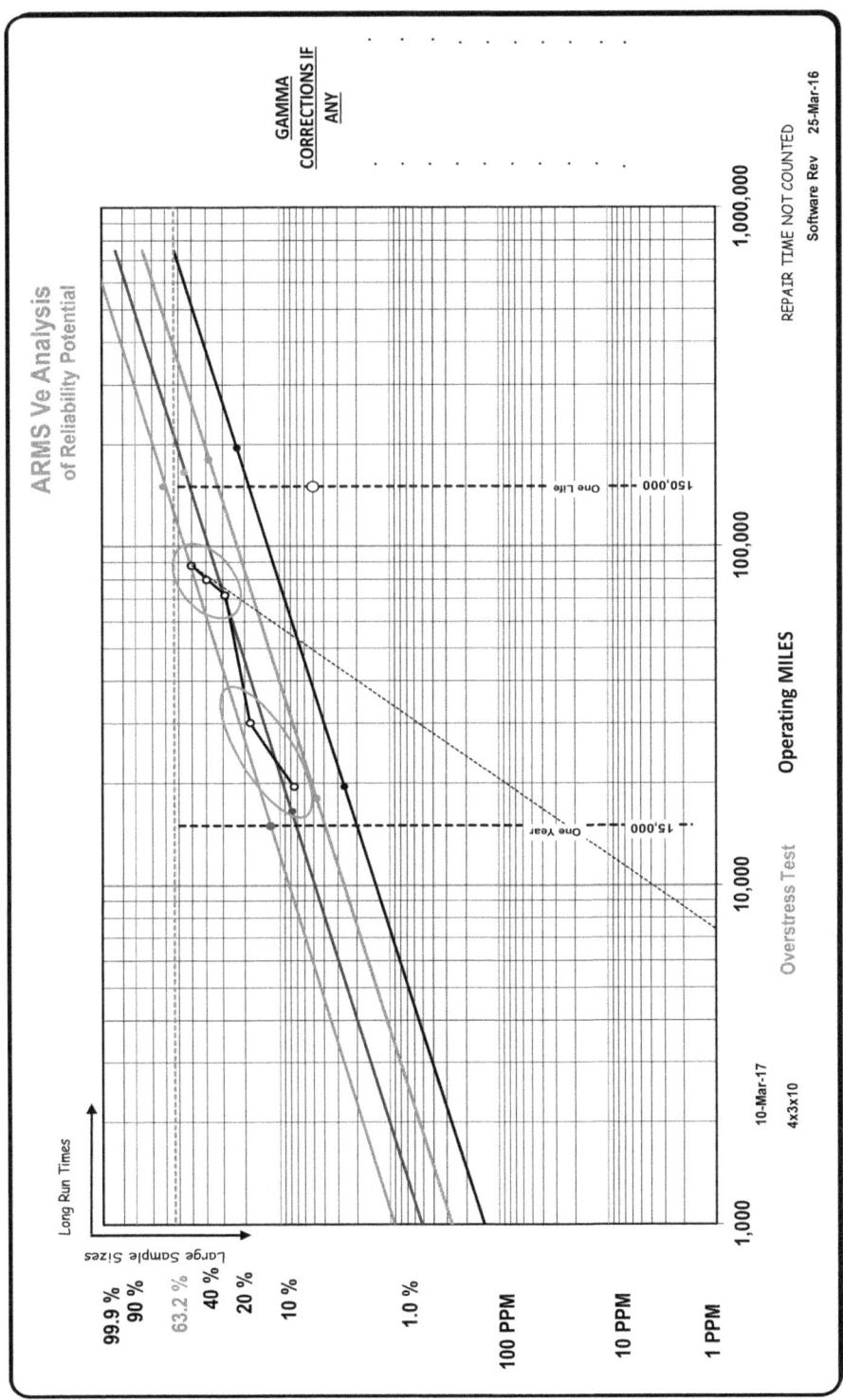

Fig. 10-9

201

The data decomposed into two clusters by visual inspection of the ARMS' Weibull graph.

TEST 1			1 Γ =		PROJECT NAME?				
TOTAL	9		1.0	Acceleration < Factor	Line color 255, 0, 0	Beta Est_1 >>	2.20	1	30,100
Line #	Qty taken Off Test	Failure=1 Suspension=0	Rank Individual TIMES or CYCLES at Test Termination	Increment	Mean Order Number	Previous Mean Order Number	Median Rank %	Plot Point Y	Plot Point X
1			MILES		1.00				
1	1	1	1.95E+04	1.0000	1.000		7.413%	-2.5637	1.95E+04
2	1	1	3.01E+04	1.0000	2.000		18.059%	-1.6136	3.01E+04
3	7							-1.6136	3.01E+04
4									

Fig. 10-10

TEST 2			1 Γ =		PROJECT NAME?				
TOTAL	7		1.0	Acceleration < Factor	Line color 0, 0, 0	Beta Est_2 >>	7.00	1	88,000
Line #	Qty taken Off Test	Failure=1 Suspension=0	Rank Individual TIMES or CYCLES at Test Termination	Increment	Mean Order Number	Previous Mean Order Number	Median Rank %	Plot Point Y	Plot Point X
2			MILES		1.00				
1	1	1	7.20E+04	1.0000	1.000		9.428%	-2.3124	7.20E+04
2	1	1	8.00E+04	1.0000	2.000		22.952%	-1.3442	8.00E+04
3	1	1	8.80E+04	1.0000	3.000		36.476%	-0.7902	8.80E+04
4	4							-0.7902	8.80E+04
5									

Fig. 10-11

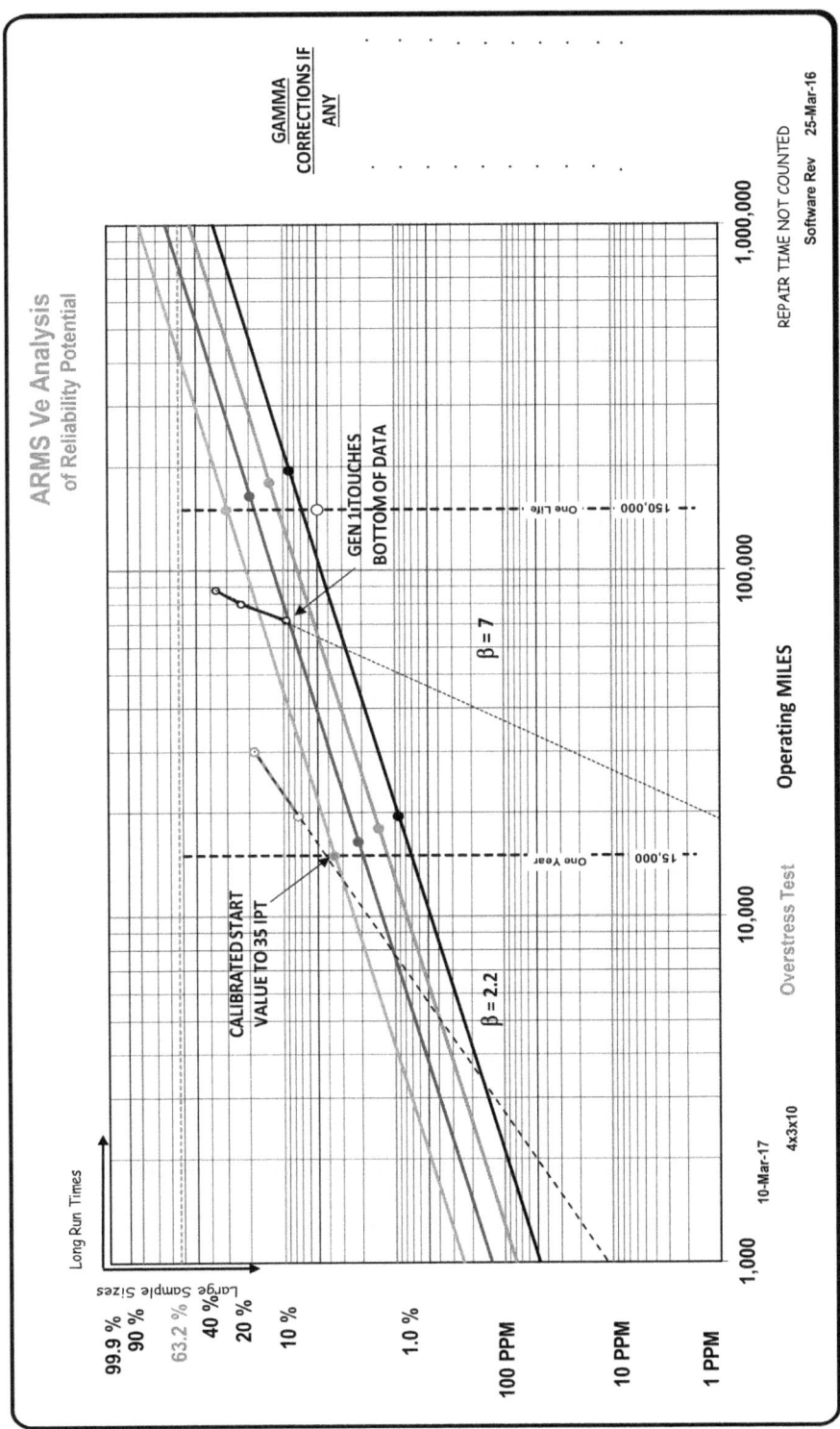

Fig. **10-12**

203

early signs of failure is required and validation testing a larger sample size is needed to capture multiple modes of failure. Based on a multitude of project timing issues and test stand schedules, the surviving units may be tested beyond the initial failure period. As we approach a production ready product, the samples are produced on production grade equipment which provide larger volumes at a lower cost. We need to take advantage of the volume to reveal significant random events and multiple failure modes.

Key questions for the Failure Analyst are why did the surviving units not fail?" and "Are there any signs of impending failure?"

Earlier, I discussed the Sub-Roc timer motors that passed all qualification tests, but were in a high state of impending failure and had to be disqualified until corrected. The same question applies to the three failures that survived conditions that caused two earlier failures. The early failures may not be due to the same cause and more than likely have separate causes. Additional work is necessary. Keeping track of build dates and critical line personnel is critical. Decomposition clarified the situation and provided direction for ongoing actions.

Recapping for Dr. Ebbinghaus: ARMS' generated Weibull lines have a slope of one, representing the bottoms of "Bath Tub" curves, and are the "max bottoms demonstrated by test." The actual bottom may be much deeper, but we have not probed the tub depth with larger samples of test units. Use caution, but develop β slopes from test data and determine if a characteristic β slope may be applied to measured data. ARMS supports entering β slope values to the data sheets by visually matching plotted results. A projection of the wearout slope estimates the potential of a deeper bathtub. To be conservative, life to the right of the projection should not be expected. This forecast is based on the median of the sample distribution.

The range of the median, based on sample size tested, can be shown by placing 1s in the cells below the "Sample Size" boxes of the Validation tab sheet. When a larger sample size is tested, the results will be projected lower in the Weibull graph and the range of the median will be less. There will also be a shift in the visual appearance of the Beta angle due to the X and Y scales being different.

Limits of the median are estimated by calculating +/- 1.645σ (95%) values of the median and plotting **exponential** values of the data, which counter the Weibull log-log vs log scales. ARMS plots the +/-95% limit values and they appear as straight lines surrounding the median. Ref. Wayne B. Nelson, "Applied Life Data Analysis" Page 131 ISBN 9780471094586. ARMS' median range limits are usually hidden during presentations to avoid data overload and confusing graphs.

This ARMS' form of representing the median range values is useful when making decisions, whereas linear values of confidence limits, as normally plotted on a Weibull graph, flare out and are virtually useless.

Testing beyond one specification life-time extends the usefulness of the samples, if they have survived to that point. Samples which have survived beyond one lifetime of testing are more valuable than untested samples. An expensive investment in time and money has already been made to test the surviving samples. At this point, a decision needs to be made concerning the additional knowledge to be gained from a detailed analysis of a surviving life test sample or from continued testing to eventual failure that may reveal new critical failure modes. There is no simple answer. The engineering team needs to review the present state of what they have learned up to that point and what they need to discover. When there are several surviving samples at the end of life testing, some may be analyzed for signs of impending new failures and the remainder put back on test. In some cases, the design may have already been replaced with a new approach. I recommend detailed analysis before scrapping samples to discover impending failure modes or to confirm knowledge of apparent success. During a very successful endurance test of Sub-Roc missile arming timers, the 100% surviving units were analyzed. These timers are electro-mechanical and contain 14 bar DC motors driving planetary gearheads which have unique recess-action gears. Post test evaluation includes timing runs at 24 VDC and 128 VDC developing 12,000 and 60,000 rpm. The two voltages are utilized to first set a window of opportunity before launch, defining when the missile would be armed, and then at the lower voltage to time how long after breaching the sea surface a nuclear warhead would detonate. Impact with the sea imparts 7000 Gs shock to the warhead. Upon examination, it was discovered that the Delrin gears at the high-speed input of the gearhead were worn-down almost completely! Only stubs of gear teeth remained on the Delrin input planetary gears. It was unbelievable that the tight timing functions had passed testing. Further analysis uncovered a small burr condition on all drive motor stainless steel gears. Corrective actions were taken to electro polish all metal gears and a new durability conformation test was successfully completed. Analysis of gear conditions following all future durability tests, pass or fail, became a new qualification requirement.

Each ARMS' model is structured with 3 generations arranged in columns right to left and having consecutive GEN #'s, or other names if you relabel them. The model will also be arranged with indentations similar to an indented drawing list. The indentions are labeled as level ONE, TWO and THREE. There are three active "Buttons" at the top of the INPUT sheet labeled ONE, TWO and THREE, which will make the respective levels visible **(Fig.5-4 & 5-5).**

The recommended way to use the ARMS' process is to conduct a GEN 1 ARMS' Workshop and obtain an estimate of the INITIAL conditions to be placed on a ARMS' Weibull graph along with a forecast generated from what the engineering team thinks they will be achieving during the next phase of development. With engineering input, the ARMS' model will display what should be expected from planned development activity during two additional design turns, GEN 2 and GEN 3. After GEN 1 is complete or approximately 3 months have passed, conduct a new ARMS' Workshop starting from scratch, but using the knowledge gained up to that point. Do not attempt to update the prior ARMS' Workshop model with add-on data. ARMS' Workshop generations are usually labeled GEN 1, Gen 2 and GEN 3 or with project specific labels. Subsequent workshops may be labeled GEN 2, GEN3 and GEN 4, if it makes sense to do so based on current progress. During each ARMS' Workshop, think of the ARMS' Weibull graph as "pinging like sonar" and providing a forecast view of future design progress. In reality, the team will be learning so much information from their current work that it would make no sense to attempt building on an historical foundation. The foundation work was as accurate as we could make it at the time, but now we know so much more and are ready to make a new forecast based on a new vision. The first ARMS' model build did provide a well thought out Pareto list of "Assignable Actions" to guide our activity. We will always be smarter in the future, even if it is to learn we were not as smart as we thought we were in the past. Developing action plans with ARMS is always better than "winging it," which is also known as flying by the "seat of our pants." During the 3-month intervals, we need to keep the team aware of any game-changing revelations so they may act accordingly.

ARMS' Weibull Decomposition: Using the ARMS' Weibull graphs to GUI decompose validation data provides valuable information on the status and progress toward isolating individual failure modes contained within a multi-failure mode design.

The following examples demonstrate how ARMS' Weibull GUI decomposition is achieved.

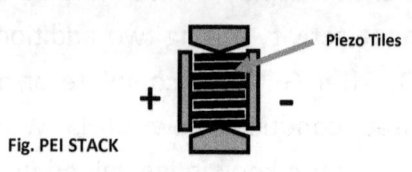

Fig. PEI STACK

Four validation tests from a Piezo Electric Injector (PEI) program provided the following data and ARMS' Weibull **(Fig.10-13,14).**

Reported cycle duration data has been modified for these examples while retaining instructional content.

A Piezo electric actuator contains a stack of over 50 ceramic tiles with alternating +/- metal sheets in between **(Fig. PEI Stack)**. When a +/- voltage is applied across each tile of the Piezo stack, each of the ceramic tiles physically expand a minuscule amount. The longer the stack the greater the total displacement. Piezo ceramic stacks create a very high force very quickly. Electric-Coil fuel injectors may cycle from open to closed 2 or 3 times during one cycle of an engine piston. A Piezo electric injector may cycle 7 to 10 times during each piston stroke while at 3500 rpm cruising speeds. This fuel metering capability results in greater fuel efficiency and lower pollution, due to having exactly the right amount of fuel required at that point in the cycle to support total fuel combustion, release of fewer hydro-carbons, greater acceleration, greater power and smoother operation, resulting in longer engine life. The promise is to attain "next generation" engines when combined with computer controls and improved advanced air throttling concepts. Several companies have applied PEI devices in a varity of configurations. External to the fuel stream and internal to the fuel stream are two major approaches, both have

Pros and Cons. This ARMS' model is for an internal type.

The validation test results **(Fig.10-13)** yielded an ARMS' Weibull

Injector Durability A2.3			γ Gamma Correction =		Program Name				
TOTAL	17		1	< Acceleration Factor		Beta Test_2 >>	0.70	1	136,739,000
Line #	QTY TAKEN OFF TEST	Failure=1 Suspension=0	Rank Individual TIMES or CYCLES at Test Termination	INCREMENT	Mean Order Number	Previous Mean Order Number	Median Rank %	Plot Point Y	Plot Point X
2			Cycles		1.00				
1	1	1	4.80E+04	1.0000	1.00	16.00	3.995%	-3.1997	48,000
2	1	1	7.20E+04	1.0000	2.00	15.00	9.746%	-2.2775	72,000
3	1	1	3.07E+05	1.0000	3.00	14.00	15.496%	-1.7815	307,000
4	1	1	3.16E+06	1.0000	4.00	13.00	21.247%	-1.4319	3,156,000
5	1	1	3.77E+06	1.0000	5.00	12.00	26.998%	-1.1562	3,773,000
6	1	1	4.87E+06	1.0000	6.00	11.00	32.748%	-0.9245	4,874,000
7	1	1	6.72E+06	1.0000	7.00	10.00	38.499%	-0.7213	6,722,000
8	1	1	1.23E+07	1.0000	8.00	9.00	44.249%	-0.5374	12,261,000
9	1	1	1.47E+07	1.0000	9.00	8.00	50.000%	-0.3665	14,675,000
10	1	1	1.93E+07	1.0000	10.00	7.00	55.751%	-0.2042	19,267,000
11	1	1	3.26E+07	1.0000	11.00	6.00	61.501%	-0.0465	32,624,000
12	1	1	3.57E+07	1.0000	12.00	5.00	67.252%	0.1100	35,718,000
13	1	1	4.60E+07	1.0000	13.00	4.00	73.002%	0.2696	45,973,000
14	1	1	4.89E+07	1.0000	14.00	3.00	78.753%	0.4376	48,944,000
15	1	1	5.54E+07	1.0000	15.00	2.00	84.504%	0.6230	55,448,000
16	1	1	5.63E+07	1.0000	16.00	1.00	90.254%	0.8451	56,335,000
17	1	1	1.37E+08	1.0000	17.00		96.005%	1.1694	136,739,000
18									

Fig. 10-13

graph having a characteristic wearout slope of 0.7 **(Fig.10-14)**. This plot pattern is not the classic bathtub shape expected and is suspected of containing multiple failure modes. There are three areas of special interest in the plotted data, and each seems to be a local area of a wearout event. They have been identified with circles on the Weibull graph **(Fig.10-14)**.

The design has several identified failure mode possibilities. The failure modes include a number of connection points, sealing systems, compression relaxation movement, piezo ceramic material contaminations, random structural defects, ceramic material composition variations, internal short circuits through and around the piezo ceramic tiles, oxygen migration in the tiles from fuel or processing fluids and other yet to be discovered failure mechanisms. An undiscovered suspect, not yet proven, is hydro-

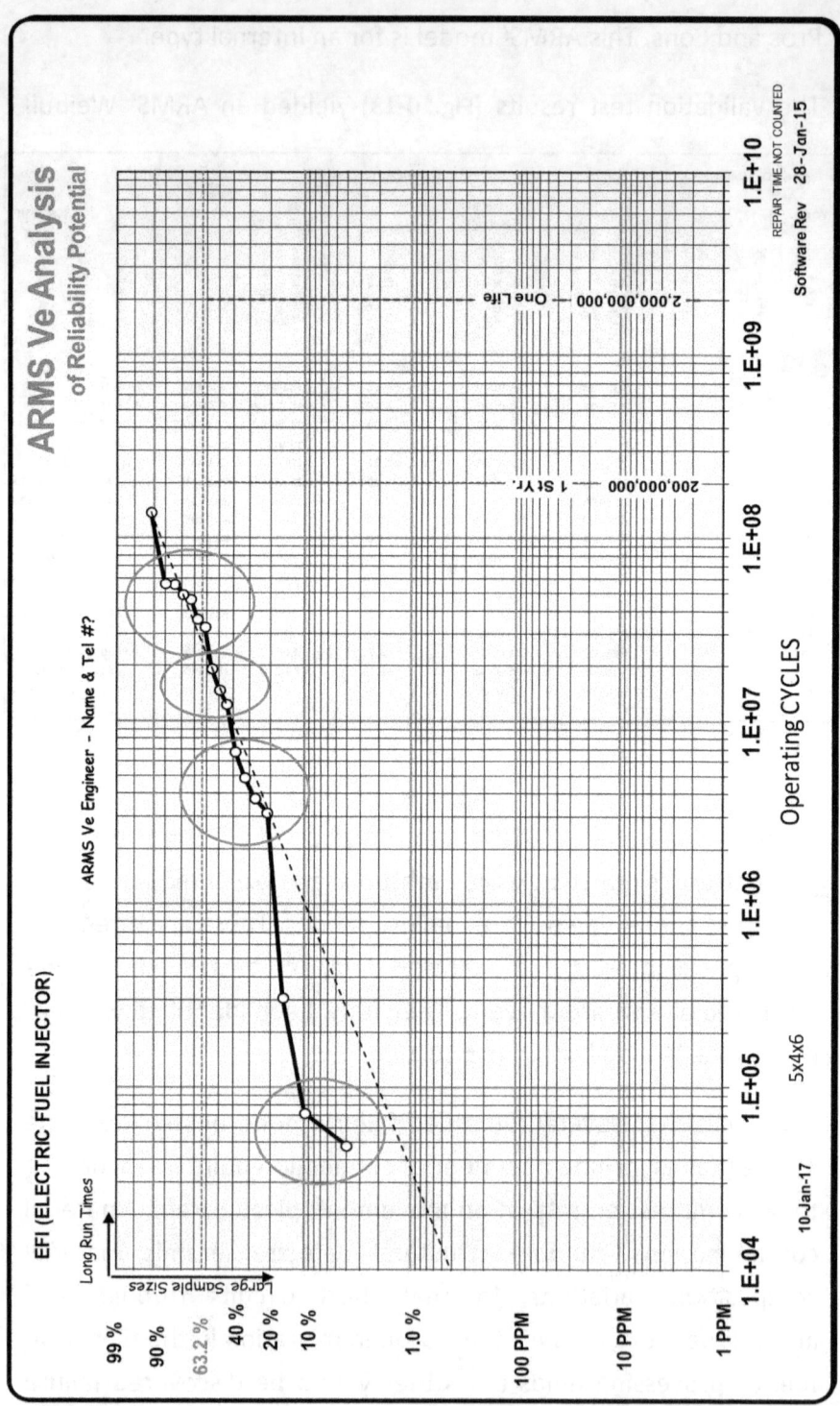

Fig. 10-14

210

gen ion migration from the fuel. I have experienced many problems in diverse products owing their failures to hydrogen ion migration. Classical hydrogen embrittlement of bolts and steel components occurs when the metal is exposed to an acid for cleaning and hydrogen ions migrate into the structure. This condition is not detectable until the component is placed under load. That is when the H− ions migrate along lines of stress to the highest point of stress where they recombine into H_2 molecules of gas and the bolt cracks from overstress. The component may actually blow apart. A similar condition occurs in hydrogen powered engines like a Sterling Thermal engine. The externally heated hydrogen transfer gas is heated to approximately +850° C and disassociates into H− ions, which can penetrate the nickel transfer tubes and be consumed. The Piezo electric injector is subject to high frequency, high voltage power at moderately high temperatures. The reason for the long explanation is to point out areas where "off line" analysis of an academic nature may be of value; design of experiments studies of sub-components or contracted studies by a university, for example. Studies may be conducted by people other than the primary engineering team to avoid getting lost in the woods.

Decomposing the PEI data into three separate sets and dropping points between sets that are not currently of interest results in an ARMS' Weibull graph **(Fig.10-15,16,17,18,19)** with four sets of data in a wearout condition. Two of the four that are close together may be with the same failure mode. As these failure modes are solved, validation test results will appear to have leaps or jumps forward. Jumps of this nature are common during innovation design and are not consistent with a continuum vision. All of the Piezo validation data plotted on the same ARMS' Weibull graph shows that multiple failure modes will cause the plot to approach a $\beta = 1$ slope unless decomposed **(Fig.10-14)**. An

Decomposed Data:

Line #	QTY TAKEN OFF TEST	Failure=1 Suspension=0	Rank Individual TIMES or CYCLES at Test Termination	INCREMENT	Mean Order Number	Previous Mean Order Number	Median Rank %	Plot Point Y	Plot Point X
TOTAL	2		1	< Acceleration Factor		Beta Test_2 >>	0.70	1	72,000
2			Cycles		1.00				
1	1	1	4.80E+04	1.0000	1.00	1.00	29.289%	-1.0597	48,000
2	1	1	7.20E+04	1.0000	2.00		70.711%	0.2053	72,000

Fig. 10-15

Line #	QTY TAKEN OFF TEST	Failure=1 Suspension=0	Rank Individual TIMES or CYCLES at Test Termination	INCREMENT	Mean Order Number	Previous Mean Order Number	Median Rank %	Plot Point Y	Plot Point X
TOTAL	4		1	< Acceleration Factor		Beta Test_2 >>	2.50	1	6,722,000
2			Cycles		1.00				
1	1	1	3.16E+06	1.0000	1.00	3.00	15.910%	-1.7528	3,156,000
2	1	1	3.77E+06	1.0000	2.00	2.00	38.637%	-0.7167	3,773,000
3	1	1	4.87E+06	1.0000	3.00	1.00	61.363%	-0.0503	4,874,000
4	1	1	6.72E+06	1.0000	4.00		84.090%	0.6088	6,722,000

Fig. 10-16

Line #	QTY TAKEN OFF TEST	Failure=1 Suspension=0	Rank Individual TIMES or CYCLES at Test Termination	INCREMENT	Mean Order Number	Previous Mean Order Number	Median Rank %	Plot Point Y	Plot Point X
TOTAL	3		1	< Acceleration Factor		Beta Test_2 >>	3.50	1	19,267,000
2			Cycles		1.00				
1	1	1	1.23E+07	1.0000	1.00	2.00	20.630%	-1.4651	12,261,000
2	1	1	1.47E+07	1.0000	2.00	1.00	50.000%	-0.3665	14,675,000
3	1	1	1.93E+07	1.0000	3.00		79.370%	0.4564	19,267,000
4									

Fig. 10-17

Line #	QTY TAKEN OFF TEST	Failure=1 Suspension=0	Rank Individual TIMES or CYCLES at Test Termination	INCREMENT	Mean Order Number	Previous Mean Order Number	Median Rank %	Plot Point Y	Plot Point X
TOTAL	6		1	< Acceleration Factor		Beta Test_2 >>	4.50	1	56,335,000
2			Cycles		1.00				
1	1	1	3.26E+07	1.0000	1.00	5.00	10.910%	-2.1583	32,624,000
2	1	1	3.57E+07	1.0000	2.00	4.00	26.546%	-1.1760	35,718,000
3	1	1	4.60E+07	1.0000	3.00	3.00	42.182%	-0.6017	45,973,000
4	1	1	4.89E+07	1.0000	4.00	2.00	57.818%	-0.1471	48,944,000
5	1	1	5.54E+07	1.0000	5.00	1.00	73.454%	0.2824	55,448,000
6	1	1	5.63E+07	1.0000	6.00		89.090%	0.7955	56,335,000
7									

Fig. 10-18

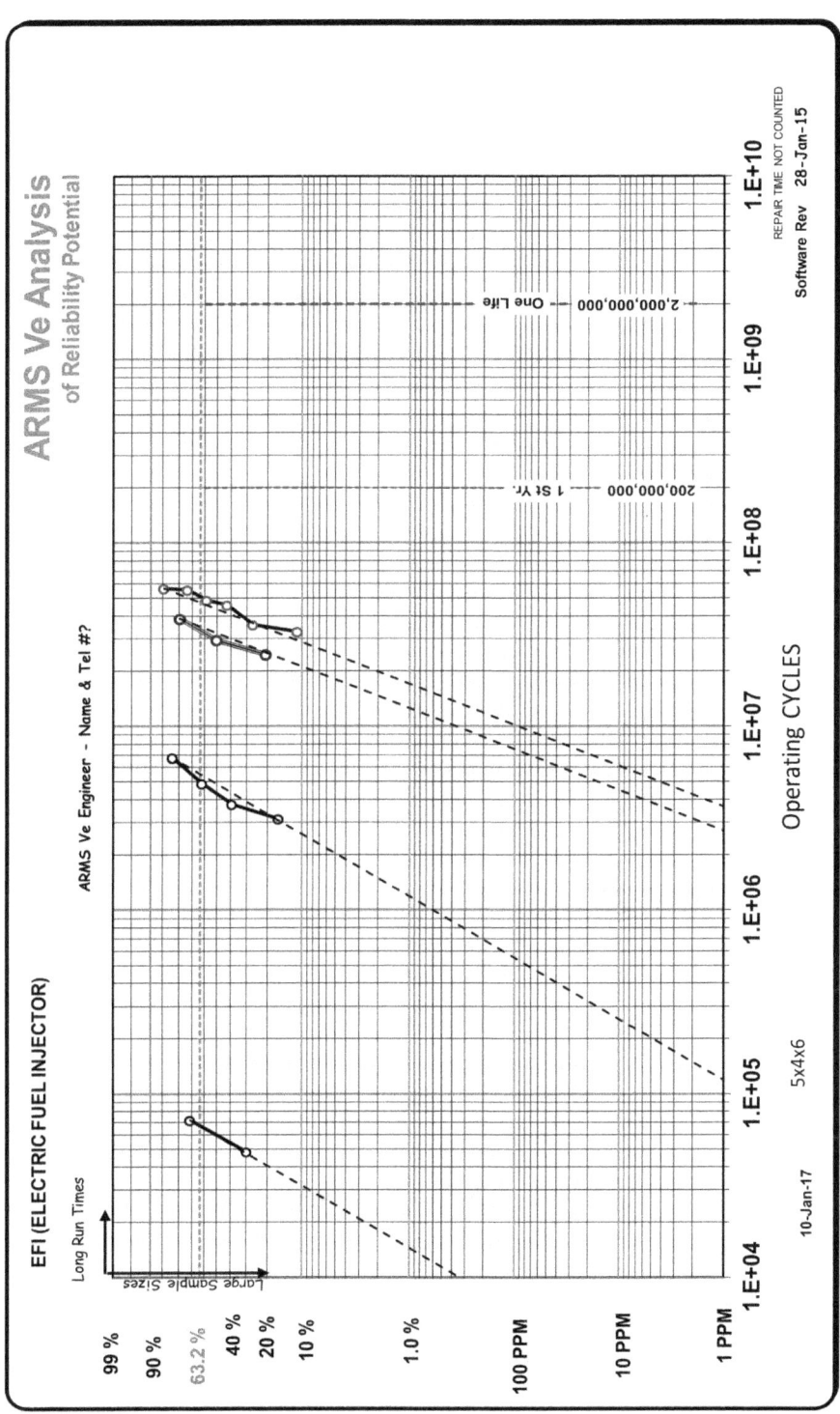

Fig. 10-19

213

ARMS' Weibull graph of the decomposed data (**Fig.10-19**) reveals apparent individual failure modes.

When test results are ARMS' Weibull plotted and appear to have a widely separated set of wearout regions, it is easy and straight forward to realize then they should be decomposed and plotted as separate sets. The separation is caused by two different modes of failure. The question should be "Why did the better set not fail earlier?" You may have a solution to the earlier problem setting right in front of you. Both are of the same design, but something made them different. Is it the build, batches, oParetors, supplier anomalies, test stand conditions?

Trying to solve anomalies of this type, I have run into some weird causes of failure. Test trials of E6 Negative Impedance Repeaters, while I was at the Bell Telephone Laboratories, were conducted in Atlanta, Georgia. The E6 Negative Impedance Repeater is a two-terminal device that has a unique ability to amplify voice signals from both directions when placed across a pair of telephone lines. To test for damage from lightning striking the telephone lines, a physical survey was made East of the Mississippi River by placing lightning hit counters on telephone lines in search of the areas most often struck by lightning. The answer was Miami, Florida, however, Atlanta, Georgia, had hills which made lightning strikes more predictable. So telephone lines were chosen that ran over hills in the Atlanta area of Buckhead and Roswell, then the E6 repeaters were installed. Everything worked fine as we waited for Mother Nature to do her job until Monday morning rolled around, when some units failed. The failed repeaters were swapped out with good units over the next few days and when Monday rolled around again, the replacements failed. This was a new design effort so the first suspect was the design or build process. After a few weeks of this action, it became clear that the problem only occurred on Mondays, mid-morning. How could this be? It turned out that lines where failures occurred ran on the same poles as power lines for several miles. On Monday morning, there was a heavy power load from people washing clothes and from manufacturers having a peak power profile as the week's work started. The A.C. magnetic field surrounding the power lines formed a transformer of one turn with the telephone lines and shifted the transistor bias of our E6 Negative Impedance Repeaters. The rest of the week and over the weekends everything was "A-OK." Solving strange patterns in your test data can require strange out of the box thinking.

A similar problem occurred in Houston, Texas. The 3D Warranty Graph shown in the Appendix **(Fig. Appx-3)** indicated that 3 Phase Air Conditioning Systems in the downtown section of Houston had a higher than normal rate of failure. As a part of the failure analysis, a trip was arranged to see the local dealer for the systems. He indicated that the failure pattern was random, except for one installation they were working to solve, but had exhausted all possible failure causes. The installation was at a Baptist Church. Every Sunday during church service, when his technicians were not working, the air conditioning unit would overload and fail due to internal safety overload protectors on the compressor motor opening. The compressor had been replaced several times. 3 phase power lines were checked during the week and met all requirements. Yet every Sunday, the same failure occurred and the customer was running out of patience. I asked if they monitored the unit throughout the weekend. They did not have the necessary recording equipment. We arranged to be on site during service the next Sunday. Sure enough, it failed. A check of the 3 phase power lines at the time of failure revealed voltages were far out of balance, and the unbalanced input to the air conditioner was tripping the Klixon overload protection system. The power company was using what is known as a "Dog Leg Ground" system for single phase power distribution. With the dog leg ground system, any new single-phase power customer is supplied power by first measuring the voltage level of each of the 3 phase legs to ground, then picking the highest one and connecting between it and ground for two wire single phase service. The dog leg ground power distribution service had been used and abused for so long that on Sunday when factories were not working and people were not running clothes washers, the three phase voltages went far out of balance from the disproportionately unloaded lines. When you have strange failure data, look for strange solutions. You may recall the B58 "Baro" set potentiometer problem discussed earlier where some pilots used their foot to kick set the ground altitude level, and

Line #	QTY TAKEN OFF TEST	Failure=1 Suspension=0	Rank Individual Measures at Test Termination	INCREMENT	Mean Order Number	Previous Mean Order Number	Median Rank %	Plot Point Y	Plot Point X
Stack 8503E			γ Gamma Correction =			Program Name			
TOTAL	7		1	< Acceleration Factor		Beta Test_3 >>	2.00	1	207,000,000
3			Cycles		1.00				
1	1	1	8.00E+05	1.0000	1.00	6.00	9.428%	-2.3124	800,000
2	1	1	2.00E+06	1.0000	2.00	5.00	22.952%	-1.3442	2,000,000
3	1	1	2.50E+07	1.0000	3.00	4.00	36.476%	-0.7902	25,000,000
4	1	1	1.02E+08	1.0000	4.00	3.00	50.000%	-0.3665	102,000,000
5	1	1	1.30E+08	1.0000	5.00	2.00	63.524%	0.0085	130,000,000
6	1	1	1.73E+08	1.0000	6.00	1.00	77.048%	0.3865	173,000,000
7	1	1	2.07E+08	1.0000	7.00		90.572%	0.8593	207,000,000

Fig. 10-20

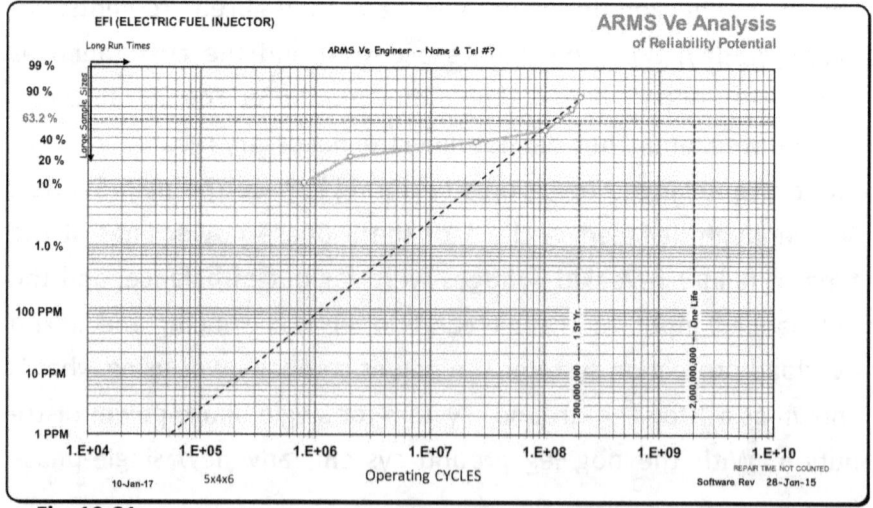

Fig. 10-21

others violently rotated them back and forth before reaching a set point. Again, strange behavior requiring out of the box solutions.

Returning to our PEI evaluations, after taking design actions to address failures revealed during failure analysis, seven PEI were tested to failure. As in the past, a flipped "S" pattern resulted. In this instance, however, there appeared to be two modes of failure.

Decomposing the seven failures into two data sets using GUI techniques **(Fig.10-23, 24)**, and graphing on an ARMS' Weibull graph yields **(Fig.10-25)**. The two sets of decomposed data appear to be in a state of wearout. I say "appears to be" because we only have two failures for the first set, and the confidence in what they are telling us is very limited. The four failures to the right side of the ARMS' Weibull graph have a β slope of 3.2 for the characteristic wearout condition. 3.2 is a reasonable value to be expected at end of life. This result is encouraging. Two failure modes appear to have been eliminated and yet we have two preliminary failures which are expected to be due to assembly errors or random causes. So again, critical failure analysis is providing clues for design development.

Injector Durability A2.3			γ Gamma Correction =		Program Name					
TOTAL	4		1	< Acceleration Factor		Beta Test_3 >>	3.20	1		207,000,000
Line #	QTY TAKEN OFF TEST	Failure=1 Suspension=0	Rank Individual Measures at Test Termination	INCREMENT	Mean Order Number	Previous Mean Order Number	Median Rank %	Plot Point Y		Plot Point X
3			Cycles		1.00					
1	1	1	1.02E+08	1.0000	1.00	3.00	15.910%	-1.7528		102,000,000
2	1	1	1.30E+08	1.0000	2.00	2.00	38.637%	-0.7167		130,000,000
3	1	1	1.73E+08	1.0000	3.00	1.00	61.363%	-0.0503		173,000,000
4	1	1	2.07E+08	1.0000	4.00		84.090%	0.6088		207,000,000
5										

Fig. 10-22

Injector Durability A2.3			γ Gamma Correction =		Program Name					
TOTAL	7		1	< Acceleration Factor		Beta Test_9 >>	1.00	1		2,000,000
Line #	QTY TAKEN OFF TEST	Failure=1 Suspension=0	Rank Individual TIMES or CYCLES at Test Termination	INCREMENT	Mean Order Number	Previous Mean Order Number	Median Rank %	Plot Point Y		Plot Point X
8			Cycles		1.00					
1	1	1	8.00E+05	1.0000	1.00	6.00	9.428%	-2.3124		800,000
2	1	1	2.00E+06	1.0000	2.00	5.00	22.952%	-1.3442		2,000,000
3	5							-1.3442		2,000,000
4										

Fig. 10-23

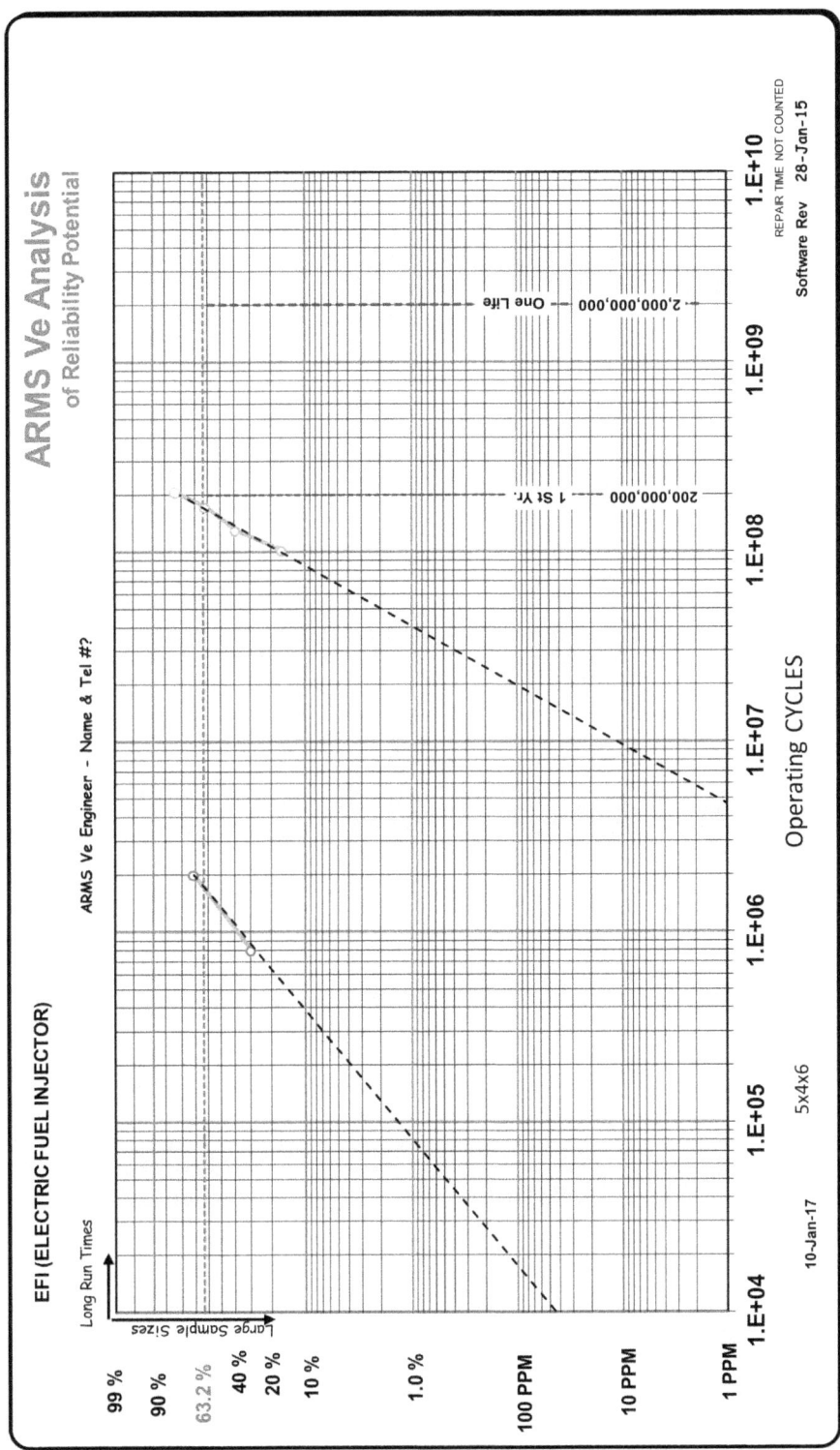

Fig. 10-24

219

The following validation test data again resulted in a flipped "S" pattern (Fig.10-25,26), however, the shape is somewhat compressed and after many more cycles of operation, it is not as clear that there are two failure modes at work.

Injector Durability A2.6			γ Gamma Correction =			Program Name				
TOTAL	4		1	< Acceleration Factor			Beta Test_2 >>	10.00	1	261,000,000
Line #	QTY TAKEN OFF TEST	Failure=1 Suspension=0	Rank Individual TIMES or CYCLES at Test Termination	INCREMENT	Mean Order Number	Previous Mean Order Number	Median Rank %	Plot Point Y	Plot Point X
4			Cycles		1.00				
1	1	1	1.62E+08	1.0000	1.00	3.00	15.910%	-1.7528	162,000,000
2	1	1	1.72E+08	1.0000	2.00	2.00	38.637%	-0.7167	172,000,000
3	1	1	2.53E+08	1.0000	3.00	1.00	61.363%	-0.0503	253,000,000
4	1	1	2.61E+08	1.0000	4.00		84.090%	0.6088	261,000,000

Fig.10-25

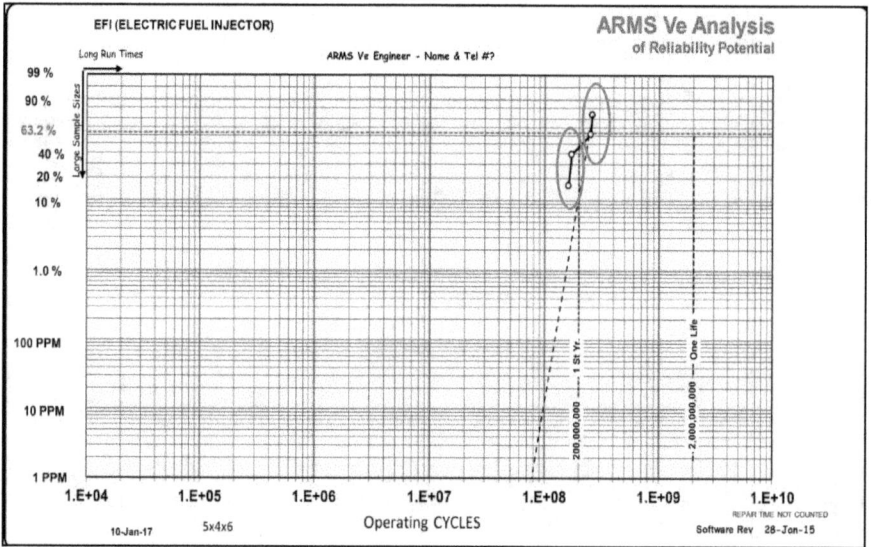

Fig. 10-26

The data was decomposed and ARMS' Weibull graphed to show separate causes. This can only be confirmed by failure analysis and/ or additional validation tests. When to display the data set as one failure mode or two modes is a judgement call.

I prefer to decompose whenever a flipped "S" pattern is present and then discuss the credibility of the approach with the design team. Often this approach brings out new revelations that the team was considering in private.

More than likely, there is one failure mode with another condition that causes an apparent delay in failure (Fig.10-25,26). The two sets could have differences in; lengths of exposure to fuel, time with voltage on without injector cycling, dwell time over weekends and a long list of other possibilities. In a separate analysis of these and several additional failures, one assembler at a supplier location was isolated as producing higher failure rates and at an earlier period in the validation test. The assembly errors were due to an oParetor wrist strength problem caused by a carpal tunnel issue. This was treated as a problem of the process rather than the assembler, and process improvements were made. Temporarily, the oParetors were retrained while the new production approach was being developed to incorporate a greater level of automation.

Decomposing **(Fig.10-26)** we obtain **(Fig.10-27,28,29)**.

Line #	QTY TAKEN OFF TEST	Failure=1 Suspension=0	Rank Individual Measures at Test Termination	INCREMENT	Mean Order Number	Previous Mean Order Number	Median Rank %	Plot Point Y	Plot Point X
Stack 8503E TOTAL 4			γ Gamma Correction = 1	< Acceleration Factor	Program Name	Beta Test_3 >> 15.00	1		172,000,000
3			Cycles		1.00				
1	1	1	1.62E+08	1.0000	1.00	3.00	15.910%	-1.7528	162,000,000
2	1	1	1.72E+08	1.0000	2.00	2.00	38.637%	-0.7167	172,000,000
3	2							-0.7167	172,000,000
4									
5									

Fig. 10-27

Line #	QTY TAKEN OFF TEST	Failure=1 Suspension=0	Rank Individual TIMES or CYCLES at Test Termination	INCREMENT	Mean Order Number	Previous Mean Order Number	Median Rank %	Plot Point Y	Plot Point X
Stack 8503E TOTAL 2			γ Gamma Correction = 1	< Acceleration Factor	Program Name	Beta Test_9 >> 25.00	1		261,000,000
8			Cycles		1.00				
1	1	1	2.53E+08	1.0000	1.00	1.00	29.289%	-1.0597	253,000,000
2	1	1	2.61E+08	1.0000	2.00		70.711%	0.2053	261,000,000
3									
4									
5									

Fig. 10-28

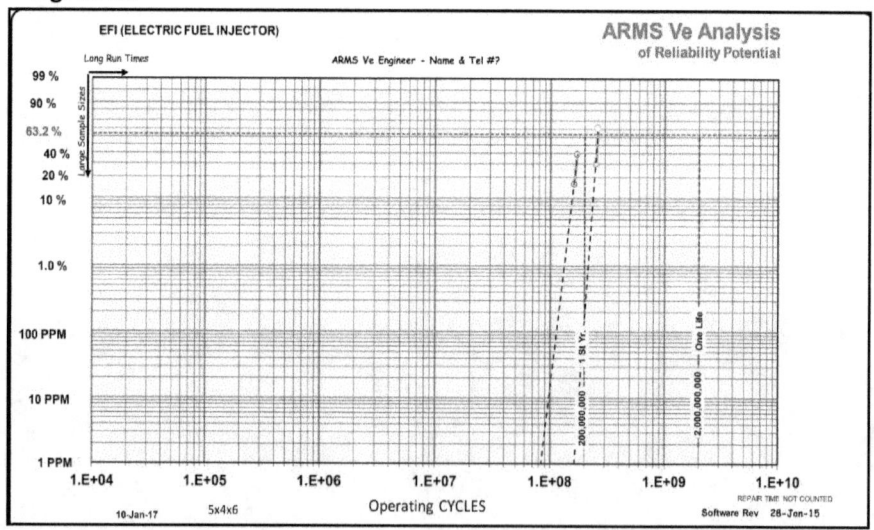

Fig.10-29

Continued design action on issues of stack sealing and connection stress relief improvement were evaluated with the next set of eleven piezo electric fuel injectors. There is general improvement in the number injection cycles, yet there is a bi-model or tri-model pattern of failure. Failure analysis is still inhibited by excessive damage during piezo material electrical shorting. Sealing is suspect. It was learned that during a ball milling operation required in preparing the ceramic materials, there was a part failure in the mill drum which introduced small metal fines into the ceramic mix. This resulted in random shorts after injection cycling.

Injector Durability A2.3			γ Gamma Correction =		Program Name				
TOTAL	11		1	< Acceleration Factor		Beta Test_4 >>	2.50	1	1,280,000,000
Line #	QTY TAKEN OFF TEST	Failure=1 Suspension=0	Rank Individual TIMES or CYCLES at Test Termination	INCREMENT	Mean Order Number	Previous Mean Order Number	Median Rank %	Plot Point Y	Plot Point X
4			Cycles		1.00				
1	1	1	1.80E+07	1.0000	1.00	10.00	6.107%	-2.7644	18,000,000
2	1	1	2.80E+07	1.0000	2.00	9.00	14.886%	-1.8253	28,000,000
3	1	1	3.90E+07	1.0000	3.00	8.00	23.664%	-1.3092	39,000,000
4	1	1	7.90E+07	1.0000	4.00	7.00	32.443%	-0.9360	79,000,000
5	1	1	2.83E+08	1.0000	5.00	6.00	41.221%	-0.6323	283,000,000
6	1	1	4.15E+08	1.0000	6.00	5.00	50.000%	-0.3665	415,000,000
7	1	1	5.70E+08	1.0000	7.00	4.00	58.779%	-0.1208	570,000,000
8	1	1	8.72E+08	1.0000	8.00	3.00	67.557%	0.1184	872,000,000
9	1	1	1.01E+09	1.0000	9.00	2.00	76.336%	0.3655	1,007,000,000
10	1	1	1.06E+09	1.0000	10.00	1.00	85.114%	0.6444	1,055,000,000
11	1	1	1.28E+09	1.0000	11.00		93.893%	1.0281	1,280,000,000
12									

Fig. 10-30

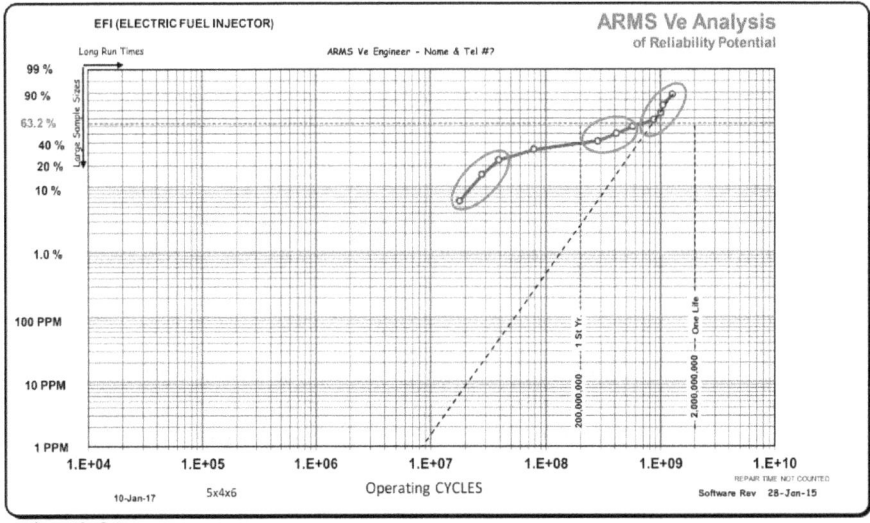

Fig. 10-31

The eleven injector results are visually decomposed into three sets of data and ARMS' Weibull graphed. This information is provided to the failure analysis team to aid in their work. Two of the first group of failures was returned to the Piezo material supplier for their analysis with specialized equipment. The metal fine condition, suspected earlier, was confirmed. Corrective actions by the supplier were in place to prevent future contamination conditions of this type. Actions to further improvements in sealing which address potential oxygen migration related failures was implemented. Improved production grade processes were required.

Injector Durability A2.3			γ Gamma Correction =		Program Name				
TOTAL	10		1	< Acceleration Factor		Beta Test_4 >>	2.00	1	39,000,000
Line #	QTY TAKEN OFF TEST	Failure=1 Suspension=0	Rank Individual TIMES or CYCLES at Test Termination	INCREMENT	Mean Order Number	Previous Mean Order Number	Median Rank %	Plot Point Y	Plot Point X
4			Cycles		1.00				
1	1	1	1.80E+07	1.0000	1.00	9.00	6.697%	-2.6691	18,000,000
2	1	1	2.80E+07	1.0000	2.00	8.00	16.320%	-1.7250	28,000,000
3	1	1	3.90E+07	1.0000	3.00	7.00	25.943%	-1.2029	39,000,000
4	7							-1.2029	39,000,000
5									

Fig. 10-32

Injector Durability A2.3			γ Gamma Correction =		Program Name				
TOTAL	7		1	< Acceleration Factor		Beta Test_9 >>	2.00	1	570,000,000
Line #	QTY TAKEN OFF TEST	Failure=1 Suspension=0	Rank Individual TIMES or CYCLES at Test Termination	INCREMENT	Mean Order Number	Previous Mean Order Number	Median Rank %	Plot Point Y	Plot Point X
9			Cycles		1.00				
1	1	1	2.83E+08	1.0000	1.00	6.00	9.428%	-2.3124	283,000,000
2	1	1	4.15E+08	1.0000	2.00	5.00	22.952%	-1.3442	415,000,000
3	1	1	5.70E+08	1.0000	3.00	4.00	36.476%	-0.7902	570,000,000
4	4							-0.7902	570,000,000
5									

Fig. 10-33

Injector Durability A2.3			γ Gamma Correction =		Program Name				
TOTAL	4		1	< Acceleration Factor		Beta Test_9 >>	5.00	1	1,280,000,000
Line #	QTY TAKEN OFF TEST	Failure=1 Suspension=0	Rank Individual TIMES or CYCLES at Test Termination	INCREMENT	Mean Order Number	Previous Mean Order Number	Median Rank %	Plot Point Y	Plot Point X
8			Cycles		1.00				
1	1	1	8.72E+08	1.0000	1.00	3.00	15.910%	-1.7528	872,000,000
2	1	1	1.01E+09	1.0000	2.00	2.00	38.637%	-0.7167	1,007,000,000
3	1	1	1.06E+09	1.0000	3.00	1.00	61.363%	-0.0503	1,055,000,000
4	1	1	1.28E+09	1.0000	4.00		84.090%	0.6088	1,280,000,000
5									

Fig. 10-34

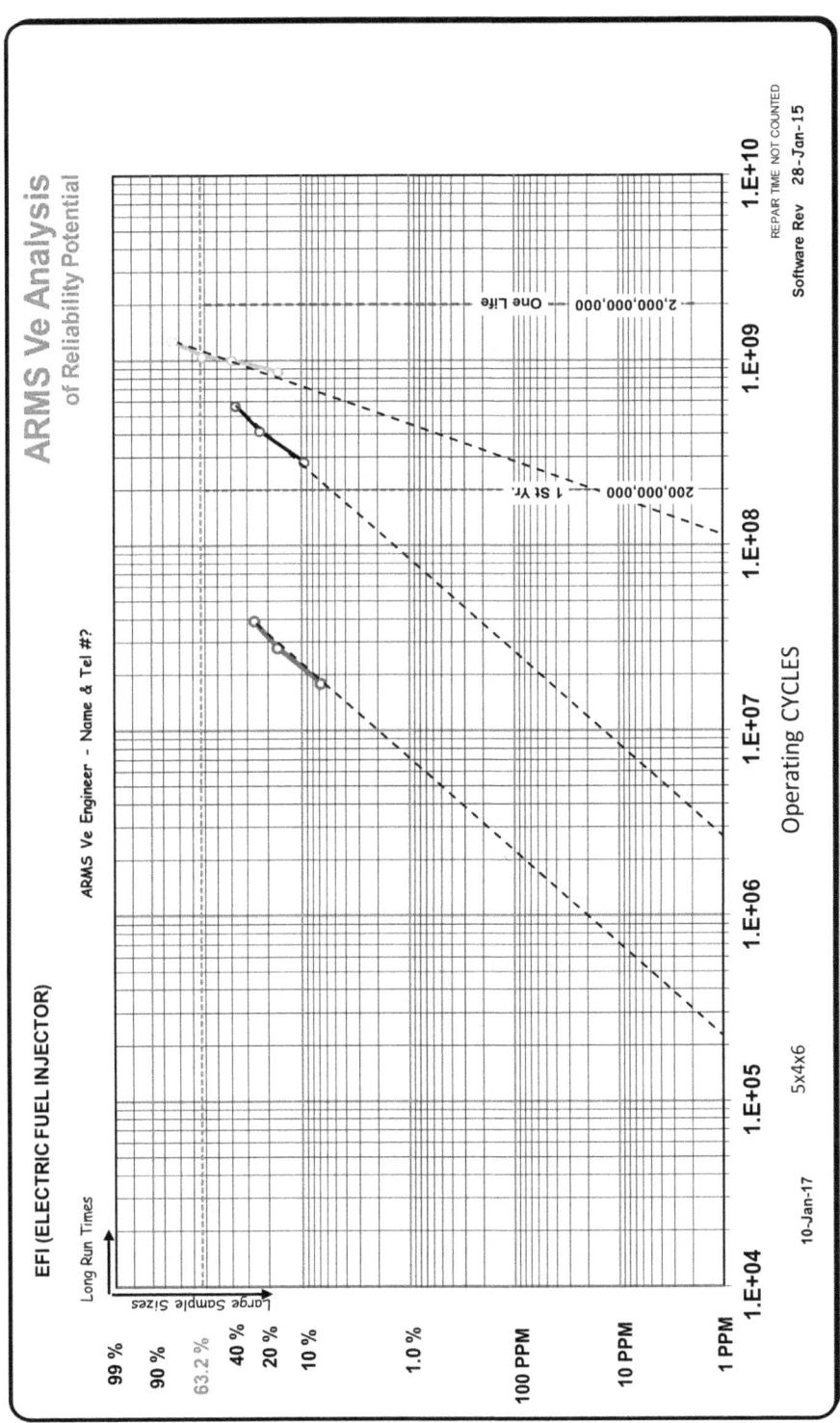

Fig. 10-35

Table 1

1			CYCLES
1	1	1	4.70E+04
2	1	1	7.30E+04
3	1	1	3.20E+05
4	1	1	3.30E+06
5	1	1	4.80E+06
6	1	1	6.80E+06
7	1	1	1.30E+07
8	1	1	1.50E+07
9	1	1	1.90E+07
10	1	1	3.30E+07
11	1	1	3.50E+07
12	1	1	6.50E+07
13	1	1	6.60E+07
14	1	1	1.40E+08

Table 2

2			CYCLES
1	1	1	4.50E+04
2	1	1	1.40E+05
3	1	1	6.20E+05
4	1	1	6.30E+06
5	1	1	7.60E+06
6	1	1	9.80E+06
7	1	1	1.30E+07
8	1	1	2.50E+07
9	1	1	2.90E+07
10	1	1	6.50E+07
11	1	1	7.20E+07
12	1	1	9.20E+07
13	1	1	9.80E+07
14	1	1	1.20E+08
15	1	1	1.30E+08
16	1	1	2.70E+08

Table 3

3			CYCLES
1	1	1	8.00E+05
2	1	1	2.00E+06
3	1	1	2.50E+07
4	1	1	1.10E+08
5	1	1	1.30E+08
6	1	1	1.80E+08
7	1	1	2.10E+08

Table 4

4			CYCLES
1	1	1	1.30E+06
2	1	1	1.40E+06
3	1	1	2.00E+06
4	1	1	4.20E+07
5	1	1	9.00E+07
6	1	1	9.20E+07
7	1	1	2.80E+08

Table 5

5			CYCLES
1	1	1	1.90E+08
2	1	1	2.80E+07
3	1	1	3.90E+07
4	1	1	8.00E+07
5	1	1	3.80E+08
6	1	1	5.80E+08
7	1	1	1.30E+09

Table 6

6			CYCLES
1	1	1	1.70E+08
2	1	1	1.80E+08
3	1	1	2.50E+08
4	1	1	2.60E+08
5	1	1	2.70E+08

Table 7

7			CYCLES
1			
2			
3			
4			
5			
6			
7			

Table 8

8			CYCLES
1	1	1	2.80E+08
2	1	1	5.60E+08
3	1	1	1.30E+09

To emphasize the contribution derived from decomposing data, seven data sets as they were reported from directly from durability stands is shown and ARMS' Weibull graphed **(Fig.10– 36,37).** This will be followed by individual decomposed data and graphs.

Fig. 10-36

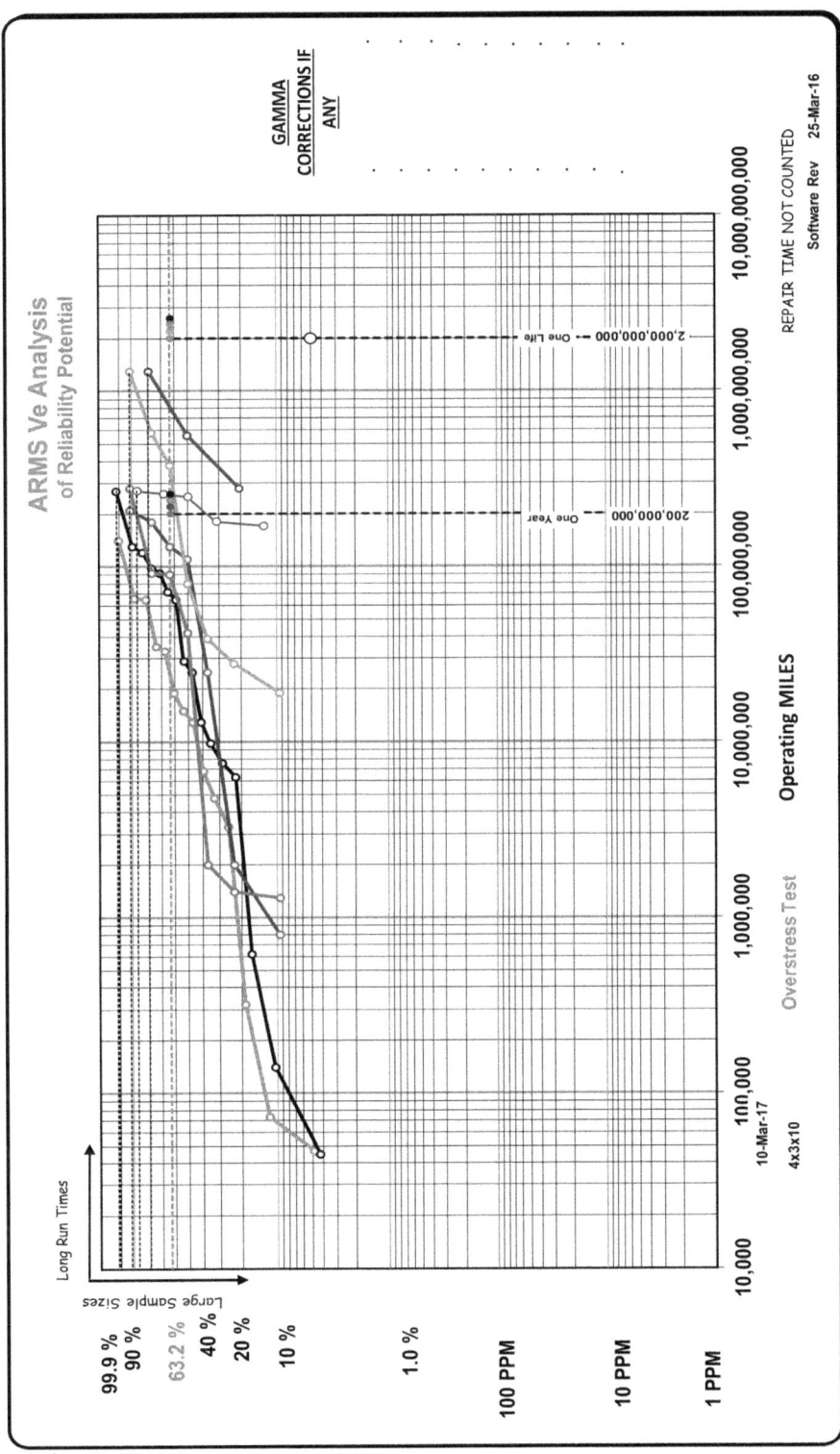

Fig. 10-37

227

1			CYCLES
1	1	1	4.70E+04
2	1	1	7.30E+04
3	1	1	3.20E+05
4	1	1	3.30E+06
5	1	1	4.80E+06
6	1	1	6.80E+06
7	1	1	1.30E+07
8	1	1	1.50E+07
9	1	1	1.90E+07
10	1	1	3.30E+07
11	1	1	3.50E+07
12	1	1	6.50E+07
13	1	1	6.60E+07
14	1	1	1.40E+08

Fig. 10-38

Validation test data from the first set of 14 Piezo Electric Injectors (PEI) reveal a pattern of failures that appear to be of a different nature (**Fig.10-38,39,40**). Decomposing the data supports this suspicion and may be used to aid in the failure analysis of the physical injectors and run data from the test program. Failure analysis did support this theory.

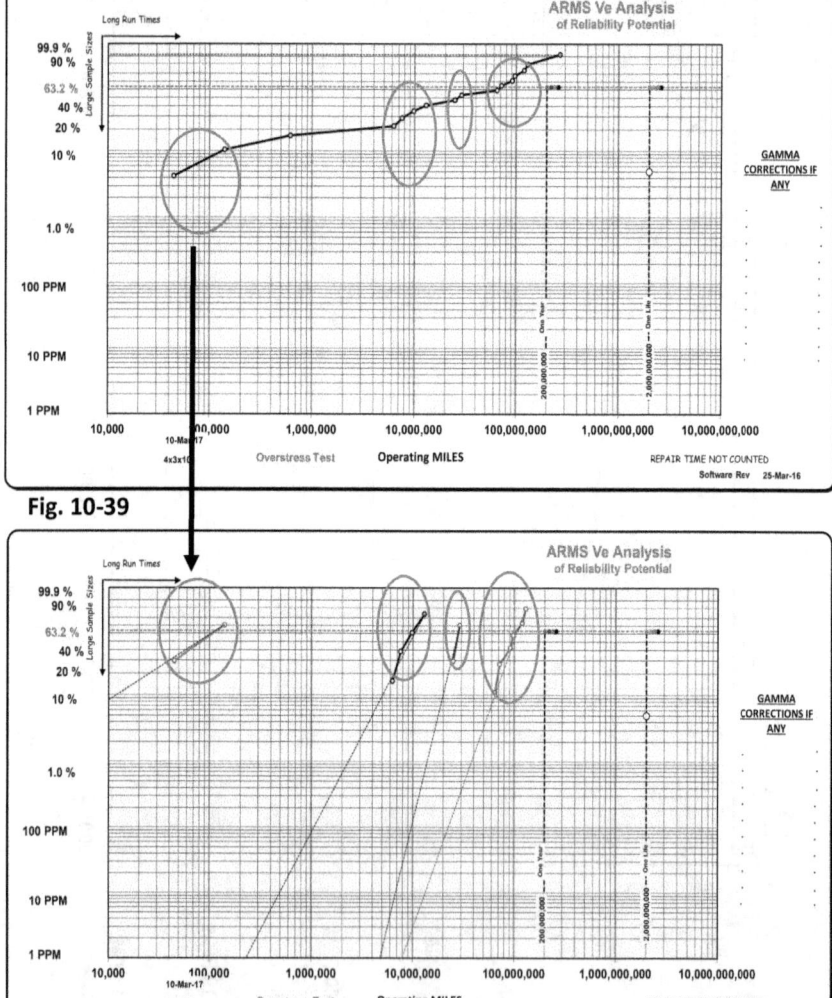

Fig. 10-39

Fig. 10-40

2			CYCLES
1	1	1	4.50E+04
2	1	1	1.40E+05
3	1	1	6.20E+05
4	1	1	6.30E+06
5	1	1	7.60E+06
6	1	1	9.80E+06
7	1	1	1.30E+07
8	1	1	2.50E+07
9	1	1	2.90E+07
10	1	1	6.50E+07
11	1	1	7.20E+07
12	1	1	9.20E+07
13	1	1	9.80E+07
14	1	1	1.20E+08
15	1	1	1.30E+08
16	1	1	2.70E+08

Fig. 10-41

This data set **(Fig.10-41)** is from Validation testing another batch of PEI made at the same time as the set shown earlier. Amazingly, the test results are nearly identical. Too identical, in fact, which raised questions about the Validation test stands being employed. Failure analysis activity revealed that the earliest failures were tested on the same validation stands.

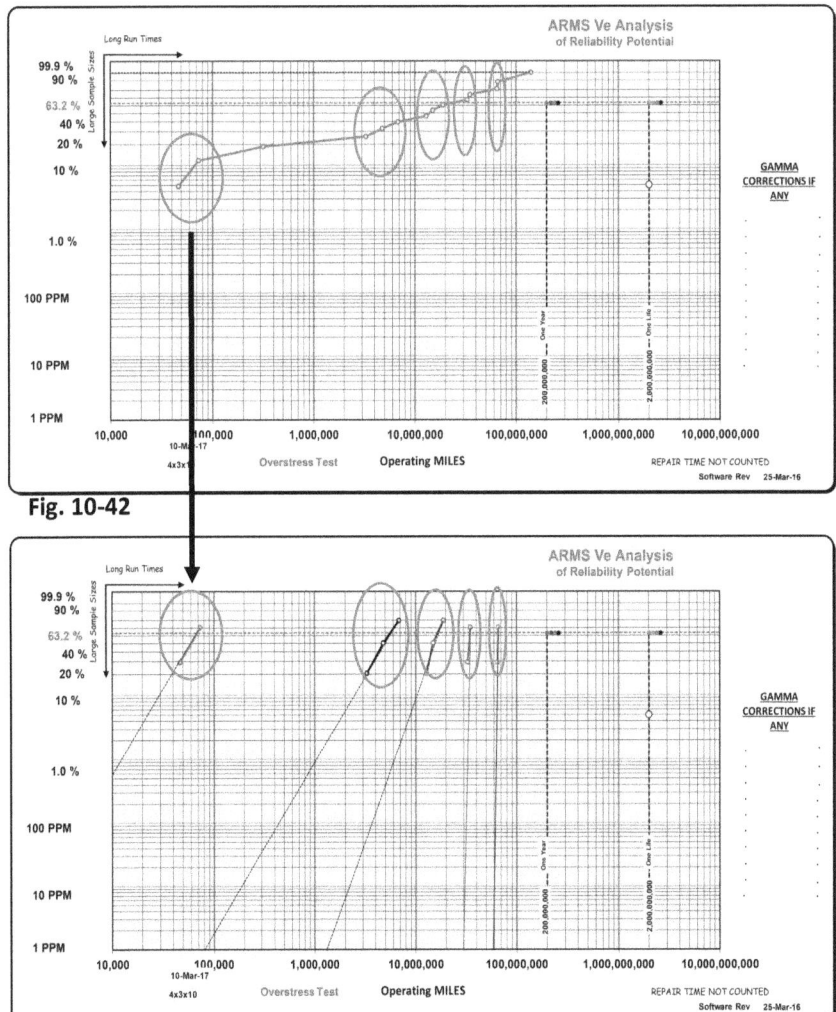

Fig. 10-42

Fig. 10-43

229

3			CYCLES
1	1	1	8.00E+05
2	1	1	2.00E+06
3	1	1	2.50E+07
4	1	1	1.10E+08
5	1	1	1.30E+08
6	1	1	1.80E+08
7	1	1	2.10E+08

Fig. 10-44

Seven new build PEI Validation tested still show Test stand and/or build anomalies with the early failures **(Fig.10-44,45,46)**. Stand hydraulic lines were re-tuned to eliminate unexpected hydraulic hammering with marginal results. Four PEI improved but are far short of meeting the 5% end of life goal. Failure analysis did reveal a specific cause of failure and corrective design actions were undertaken.

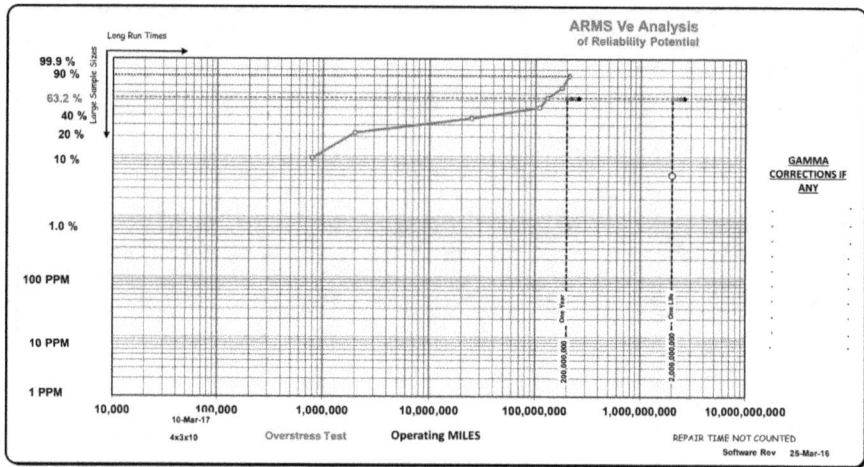

Fig. 10-45

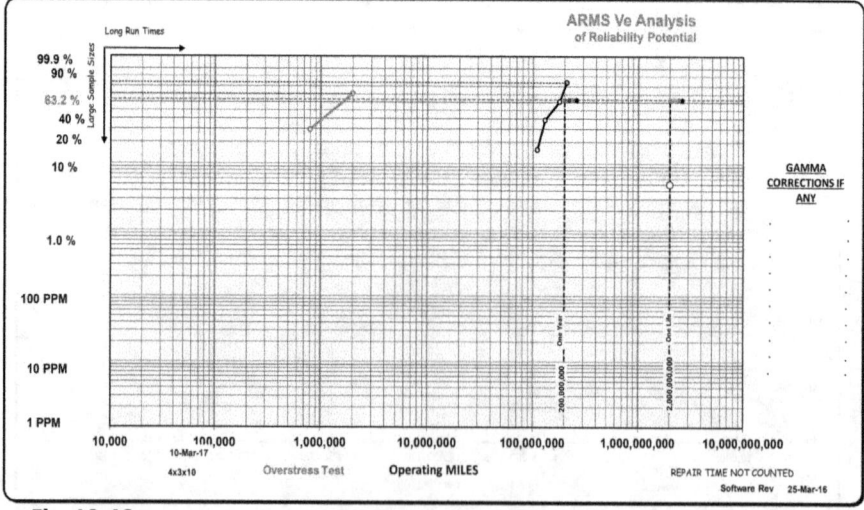

Fig. 10-46

230

4			CYCLES
1	1	1	1.30E+06
2	1	1	1.40E+06
3	1	1	2.00E+06
4	1	1	4.20E+07
5	1	1	9.00E+07
6	1	1	9.20E+07
7	1	1	2.80E+08

Fig. 10-47

Seven additional units **(Fig.10-47)** from the same build produced similar validation test results. While the Validation stand hydraulics are still considered an important factor for early failures, they are not the only factor. Certain build parameters have been identified as engineering build oParetor sensitive. During the next ARMS' Workshop, a Support Factor (SF) was opened to obtain production equipment on line earlier than planned.

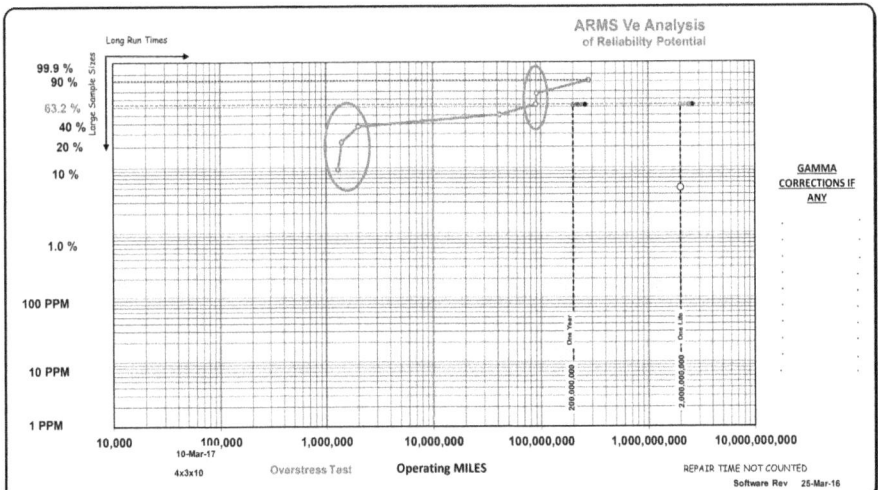

Fig. 10-48

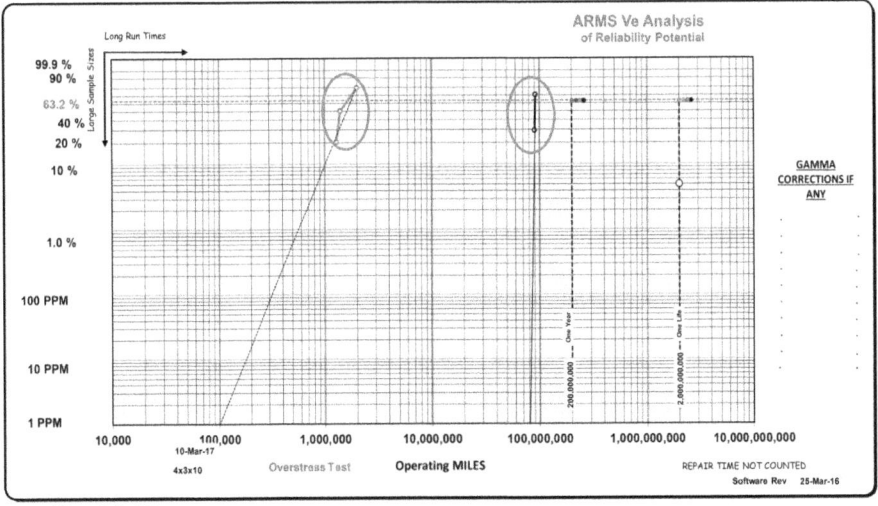

Fig. 10-49

5			CYCLES
1	1	1	1.90E+07
2	1	1	2.80E+07
3	1	1	3.90E+07
4	1	1	8.00E+07
5	1	1	3.80E+08
6	1	1	5.80E+08
7	1	1	1.30E+09

Fig. 10-50

Design improvements centered on process changes to facilitate assembly of sub-components and final assembly on the engineering assembly line. All processes were reviewed and oParetors reinstructed. Certain basic design elements were changed based on failure analysis. The test stands were equipped with current limiting devices to limit the degree of destruction during failures. Limiting destruction would aid in the analysis of failures. It was obvious at this point in the development that there was too much "craftsmanship" required on the engineering line. Greater Production Engineering involvement was required (Fig. 10-50,51,52).

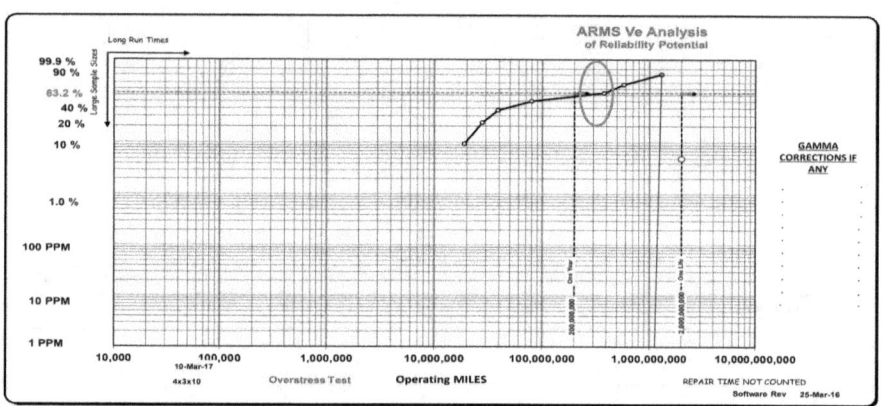

Fig. 10-51

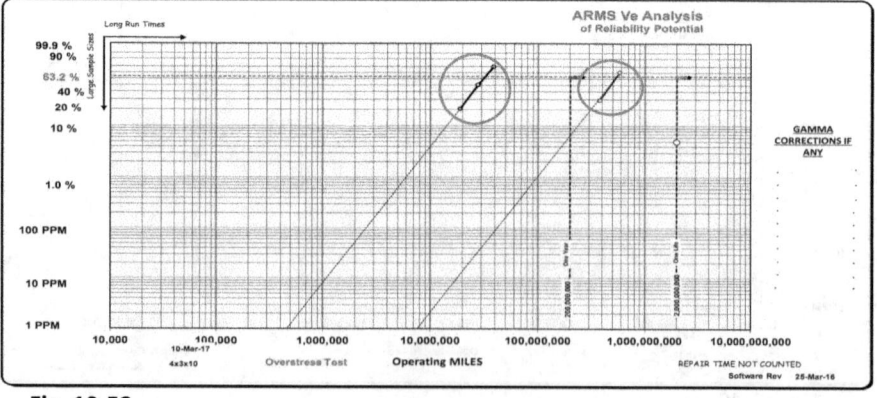

Fig. 10-52

6			CYCLES
1	1	1	1.70E+08
2	1	1	1.80E+08
3	1	1	2.50E+08
4	1	1	2.60E+08
5	1	1	2.70E+08

Fig. 10-53

Greater utilization of production grade equipment appears to have improved results. It is not clear that the failures are bi-model. The data is shown decomposed based on the shift in results **(Fig. 10-53,54,55).** Failure analysis is required to clarify results. PEI samples are now being produced in greater volume as production-ized assembly is being developed. Elements of required "Craftsmanship" have been removed. Stack sealing is receiving improved development actions. Connection stress relief and stack compression are more consistent at design levels. Additional improvements in piezo chemistry have improved supplier yields and stack performance. A great deal more development work is required and under way to meet the goal.

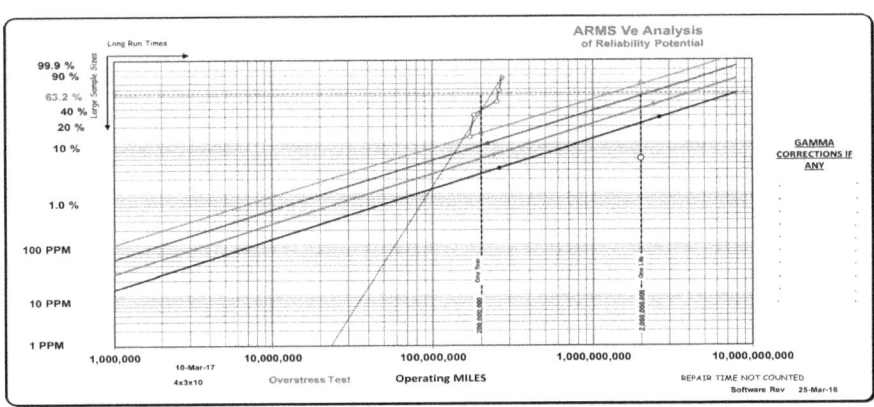

Fig. 10-54

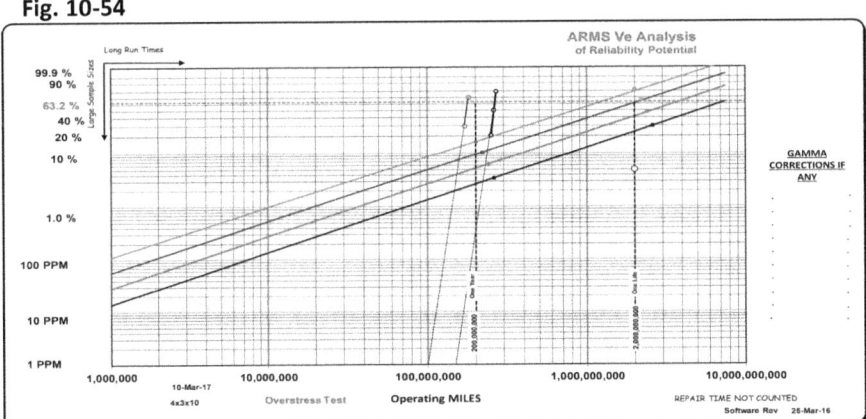

Fig. 10-55

8			CYCLES
1	1	1	2.80E+08
2	1	1	5.60E+08
3	1	1	1.30E+09

Fig. 10-56

After some minor adjustments, three additional PEI units from a new build were tested. The slope appears to be approaching a slope of one, however, it will take a much larger sample to confirm this theory and to support Weibull' plots lower on the graph **(Fig.10-56,57)**. Progress is clear, however, it

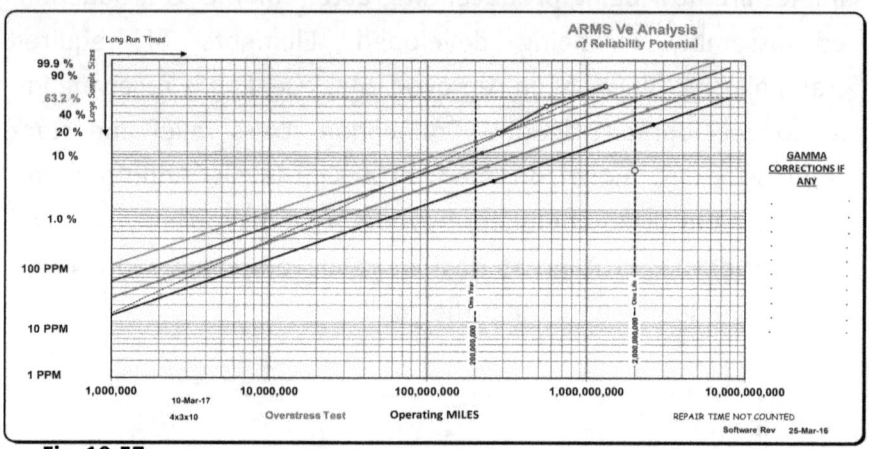

Fig. 10-57

will require 6 to 8 additional cycles at a minimum to reach the required goal.

Very many design changes were made early in development, avoiding what would have been expensive changes during Production Development. To continue will require less expensive production produced PEI at greater volumes to support the sample size trials that would be required. Additional customer engine test stands are planned to increase test capability. At this juncture, the responsibility for development transferred to the Production Engineering group for continued development. The change was normal for the company's development plan. There were many administrative issues beyond test results that would be resolved with the change. There were many technical issues discovered up to this point that cannot be discussed in detail for proprietary reasons, issues such as with connections, stress relief, containment pressure application, piezo material physics and chemistry, Piezo stack structure and

234

grinding techniques, stack sealing and chemical interactions when installed in the customer's environment.

Clearly decomposing the data reported by the durability test group was required to clarify what was occurring as the design was being improved. It is also clear that if multiple failure modes are involved, it is advisable to increase the sample size in order to have adequate sample sizes in the decomposed data.

This project was supported by engineers participating in ARMS' Workshops at several locations across Europe and the United States of America involving principal engineers and supplier personnel during a three-year period. Most ARMS' Workshops were at the request of a local engineering team to help reorganize and prioritize their work as they sought breakthroughs and incremental steps of progress. It was a pleasure to work with the teams and to see their efforts move on to the next level of development. The Advanced Engineering Group had new challenges to consider prioritizing and on which to focus their attention.

Chapter Eleven

Extended Life & Overstress Testing

- Always include normal conditions in your test program as well as perceived overstress conditions.

- Facilitating Outstanding Leadership / Teamwork between all departments

- Engineering Design & Innovation Under Control

ARMS does not tell you how to setup and conduct overstress or extended life tests, however, ARMS has provisions for accommodating tests of this type. The ARMS' Weibull graph displays a DOT for the maximum time, cycles or distance that test samples are planned to be tested, if they do not experience failure. These DOTS can be labeled with the data point values for the maximum extended test planned if you wish.

On the data INPUT sheet, there are GEN 1, 2 and 3 boxes for "Test in Life Times." If the plan is to extend testing an additional 20% longer, the value entered would be 1.2 lifetimes. The example displayed probes to the right with an increasing value for extended life at each design generation **(Fig. 11-6)**.

Testing to failure is highly recommended, but test stand availability may dictate suspending tests in order to expand the sample size or accommodate newer design samples waiting for test. If extended testing is not planned, the input data value is set to one. At the end of a scheduled test period, units still running may be reported as suspended. ARMS can display "what if" results to aide in determining when it makes sense to run longer or when to suspend the remaining samples.

As shown **(Fig. 11-2,3)**, the planned test times, cycles or distances will be indicated on the associated Weibull graph.

Overstress testing is a much more complex challenge than extended life testing. To setup an overstress test program requires more intimate knowledge of the product and intended applications.

Stress levels during normal application need to be developed. Tools such as computer aided design and finite element analysis can provide tremendous support.

During development of the B58 "Hustler" aircraft, it was decided that the B58 would be a "Demonstrated Reliability Program." This required validation testing beyond normal life while at worst case combined environments. Special thermal chambers were built to test aircraft components and the flight instruments. I was developing a new BDHI (Bearing Degree Heading Indicator) for the aircraft. Our test chamber was a hi/lo temperature chamber (+125C to -55C) that had the ability to thermally shock the test units by rapid temperature changes. At the same time, while at temperature, test units were mounted on a shaker table that could be set for sine wave or random vibrations. Mechanical shocks were built into the shaker table that seemed like we were hitting the table with a big hammer. Additionally, to test for any moisture internal to the BDHI, we conducted a "fog test." A fog test requires stabilizing an instrument at the max temperature, 125C, and rubbing the glass screen with an ice cube to see if any moisture would condensate on the interior side of the glass. Eventually, we learned how to survive all of these tests.

A hand set control knob was on the front of the BDHI to set local barometer readings. This is necessary for altitude calibration. The BDHI is an instrument critical to aircraft "flight safety," because it is the aircraft's compass and altimeter. It was decided to turn the hand set knob with an external stepper motor at ten times the maximum speed considered normal when turning by hand, and to also snap from CW to CCW across the full range of motion. Durability test acceptance was set at 100 times normal life. Even under these harsh conditions, no problems were encountered during development and the BDHI was deemed ready for deployment. After a few months in service, failures of the knob started to occur. Units re-

IPT Start	100	33	PROCESS % SIGMA				
ONE YEAR TIME	440	5,500	ONE YEAR MILES				
ONE LIFE in YEARS	12.5	1.0	OVERSTRESS FACTOR	1.0	TEST OVERSTRESS	1.0	TEST OVERSTRESS
ONE LIFE GOAL	5.0%	1.0	TEST IN LIFETIMES	1.0	TEST IN LIFETIMES	1.0	TEST IN LIFETIMES
ONE LIFE DISTANCE	68,750	6	SAMPLE SIZE	9	SAMPLE SIZE	12	SAMPLE SIZE
	GEN 1			GEN 2		GEN 3	

Fig. 11-1

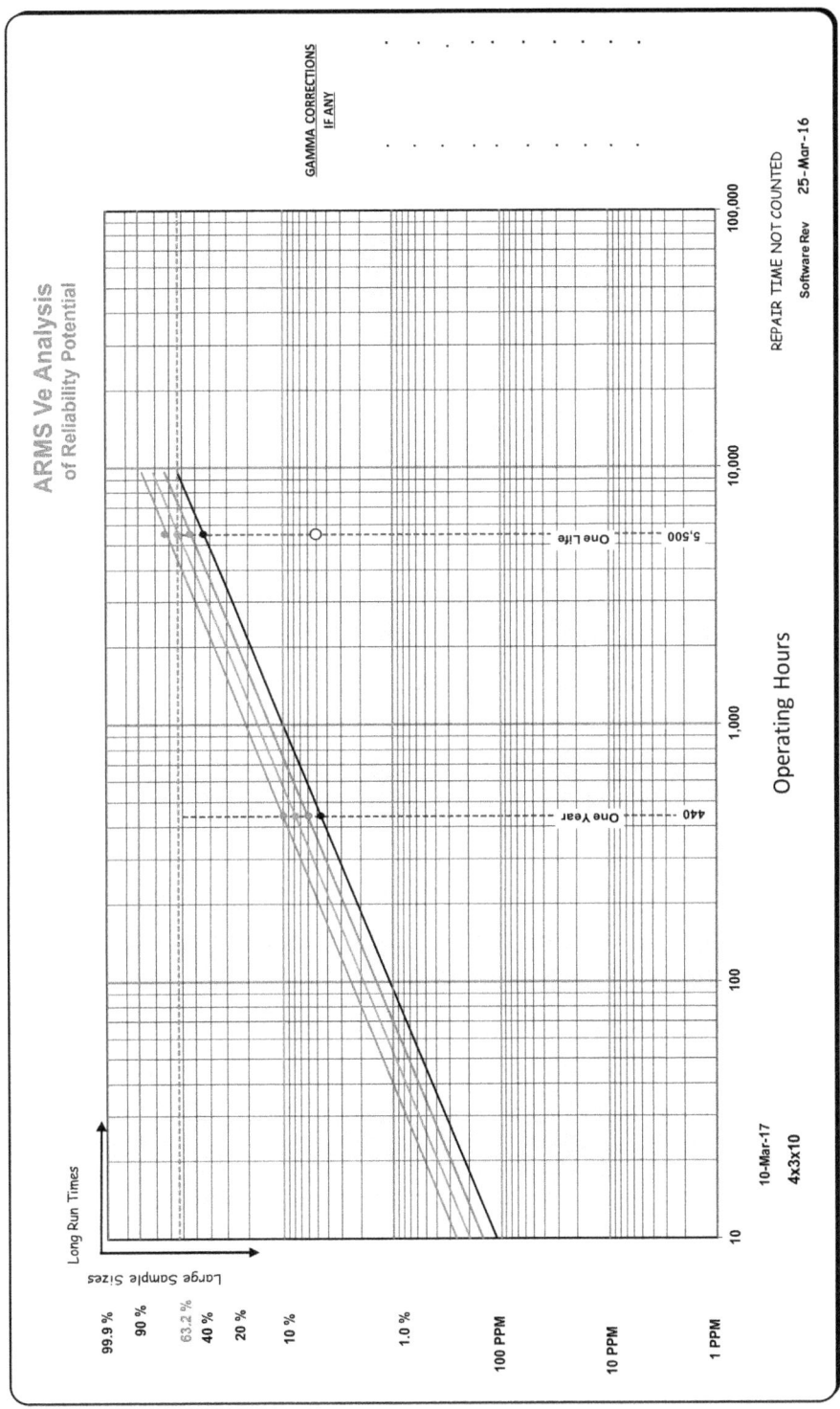

Fig. 11-2

239

turned were analyzed and the wire wound potentiometer element, on which a sliding contact wiper turned, was worn through, element wires were actually broken. How could that be? The usage was a fraction of the expected life. The qualification program was revamped to triple the run time and increase the rotational speed in an effort to force failures. Test failures did not occur. The problem was perplexing. To gain additional information, I went to an Air Force base near Dallas, Texas, to talk to the B58 pilots and flight maintenance crews, a really nice bunch of highly qualified guys. The problem eluded solution until one of the pilots mentioned that the new B58 airplane was difficult to fly because they were tightly confined in their seat and wearing parachutes, which restricted movements. The easiest way to adjust the BDHI was to reach up with your foot and basically give it a little kick! We had no test of that nature in our test plan! During simulations back in the lab, it was noted that the side pressure could put additional pressure on the wiper arm and wire wound element. The potentiometer shaft and sleeve bearing were redesigned to tolerate the added side pressure and new qualification tests were run with 50% higher side pressures simulating reported "kicks." All sample units passed qualification and the new design BDHI instruments were raced to the field for installation. (Notice the pattern of no failures found!) A Validation Test Plan must be designed to produce failures. Within a month, the new BDHI's were failing again! After brainstorming and a few sleepless nights, we decided to test by duplicating actual application conditions instead of overstress accelerated conditions. The test stands were slowed down and a dwell was added at the ends of each rotation. Failures occurred rapidly and duplicated field application results! Failure analysis revealed that the failures were due to wear during low speed "static" lubrication conditions between the potentiometer element and wiper contact. Prior testing at increased speed and high side loading was developing and maintaining a "dynamic" lubrication film on the wire wound element and no failures. Slow movement with static lubrication and a dwell at the ends of the rotation allowed the lubrication film to be penetrated by the wiper element and directly, metal to metal, rapidly wear.

Always include normal conditions in your test program as well as perceived overstress conditions.

In this case, overstress testing was actually lower stress! Static vs dynamic lubrication wear is a fundamental failure mechanism I have since encountered on several projects.

Parallels can be found in electronic tests. IGBT Power Switches that switch at the wrong times and DC motor commutation currents internal to the motor that are not detectable on the power leads are two cases that come to mind. The IGBTs stopped BART commuter trains in San Francisco, Diesel electric locomotives in Baltimore and Hybrid Busses in L.A. DC motor internal currents became an issue of national security when they delayed the KH11 Reconnaissance Satellite and the Viking Mission to Mars. Designs which have internal energy stored in inductive coils, capacitors, heatsinks or flywheels, when the system is turned off, need to have a safe method to discharge the energy. These seemingly "minor" conditions can become very important.

With overstress testing, the ARMS' Weibull lines spread while the test clock time remains the same as with no overstress.

Run time changes with extended life testing. The DOTS represent how long we plan to run the Durability Test Stand and the ARMS' Weibull lines show the effect of an extended test time.

If the sample units can be kept on test, you will demonstrate greater reliability than when stopping at one life time under normal usage stress. If the units fail, plot the failure data and recali-

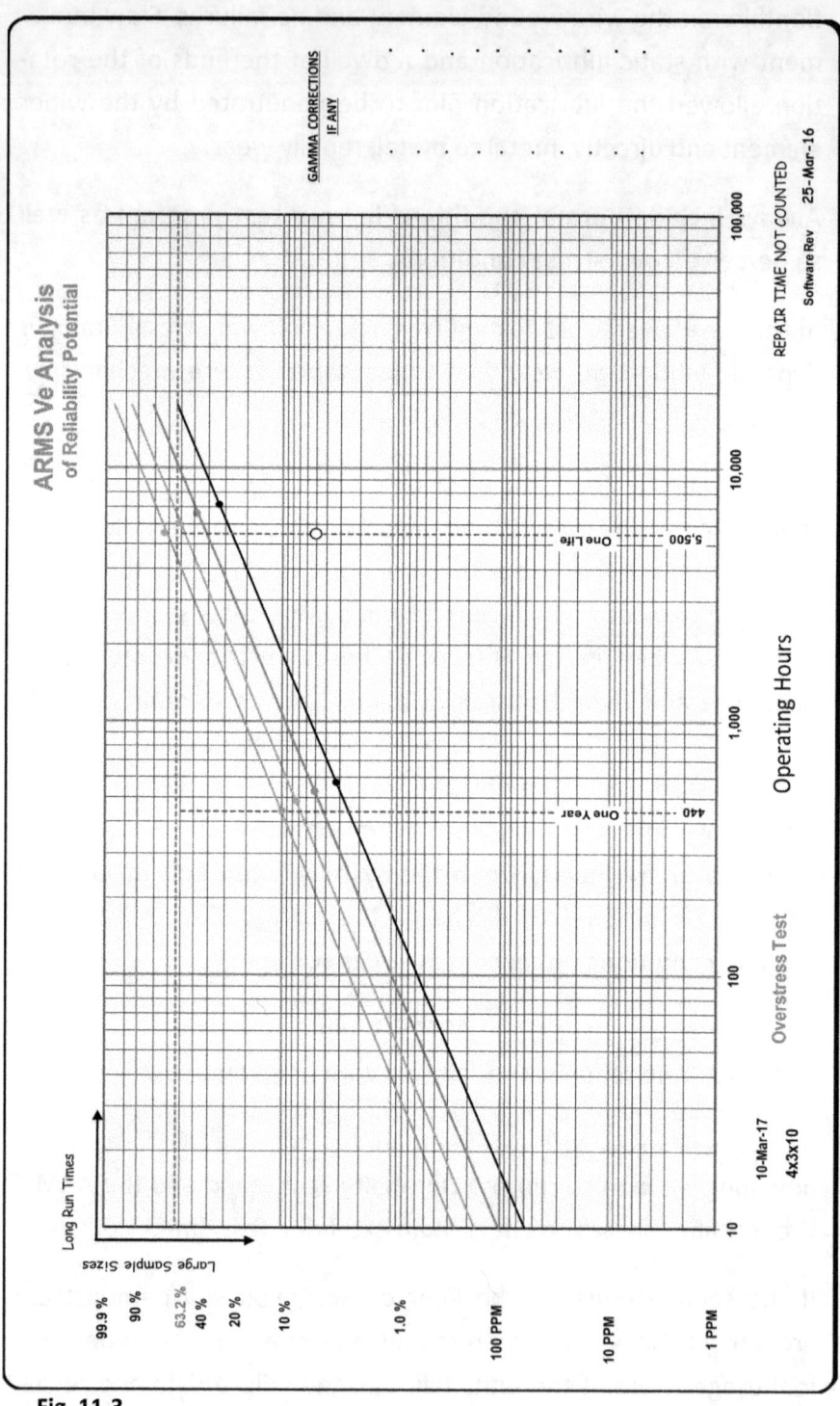

Fig. 11-3

IPT Start I0	100	25	PROCESS % SIGMA				
ONE YEAR TIME	500	15,000	ONE YEAR MILES				
END of LIFE YEARS	10.0	1.0	OVERSTRESS FACTOR	1.0	TEST OVERSTRESS	1.0	TEST OVERSTRESS
END of LIFE GOAL	1.0%	1.0	TEST IN LIFETIMES	1.0	TEST IN LIFETIMES	1.0	TEST IN LIFETIMES
END of LIFE DISTANCE	150,000	6	SAMPLE SIZ	9	SAMPLE SIZE	12	SAMPLE SIZE

Fig. 11-4

IPT Start I0	100	25	PROCESS % SIGMA				
ONE YEAR TIME	500	15,000	ONE YEAR MILES				
END of LIFE YEARS	10.0	1.2	OVERSTRESS FACTOR	1.3	TEST OVERSTRESS	1.4	TEST OVERSTRESS
END of LIFE GOAL	1.0%	1.0	TEST IN LIFETIMES	1.0	TEST IN LIFETIMES	1.0	TEST IN LIFETIMES
END of LIFE DISTANCE	150,000	6	SAMPLE SIZ	9	SAMPLE SIZE	12	SAMPLE SIZE

Fig. 11-5

brate the ARMS' Weibull lines by adjusting the INPUT start value.

Bay Area Rapid Transit (BART) hybrid rail cars, mentioned earlier, were plagued with the failures of IGBT power electronics. It was finally determined that underload, the input power rectification from positive to negative, was not in synchronization with the A.C. input wave form, and was occurring at a point where the voltage was at a non-zero value. The I^2R heating that resulted from current surges during this brief switching period led to premature IGBT failure. This had not been anticipated during validation testing because units were overstressed, and failures were attributed to overstress rather than a built-in stress due to the switching anomaly underload. At

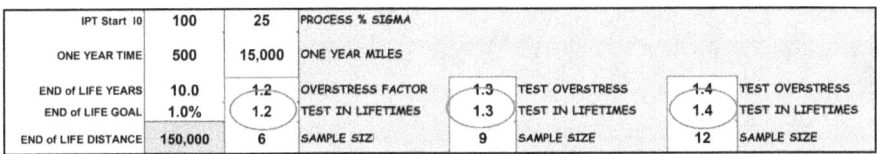

IPT Start I0	100	25	PROCESS % SIGMA				
ONE YEAR TIME	500	15,000	ONE YEAR MILES				
END of LIFE YEARS	10.0	1.2	OVERSTRESS FACTOR	1.3	TEST OVERSTRESS	1.4	TEST OVERSTRESS
END of LIFE GOAL	1.0%	1.2	TEST IN LIFETIMES	1.3	TEST IN LIFETIMES	1.4	TEST IN LIFETIMES
END of LIFE DISTANCE	150,000	6	SAMPLE SIZ	9	SAMPLE SIZE	12	SAMPLE SIZE

Fig. 11-6

the time, power IGBTs were a new item and the engineering community was still learning how to use them. A cause for the shift in the switching point was a very large foil wrapped capacitor with the foil wraps moving due to higher voltage forces after installation. At high load, the greater non-zero switching currents and resultant heating were also higher.

ARMS provides the means to enter an overstress factor. The factor

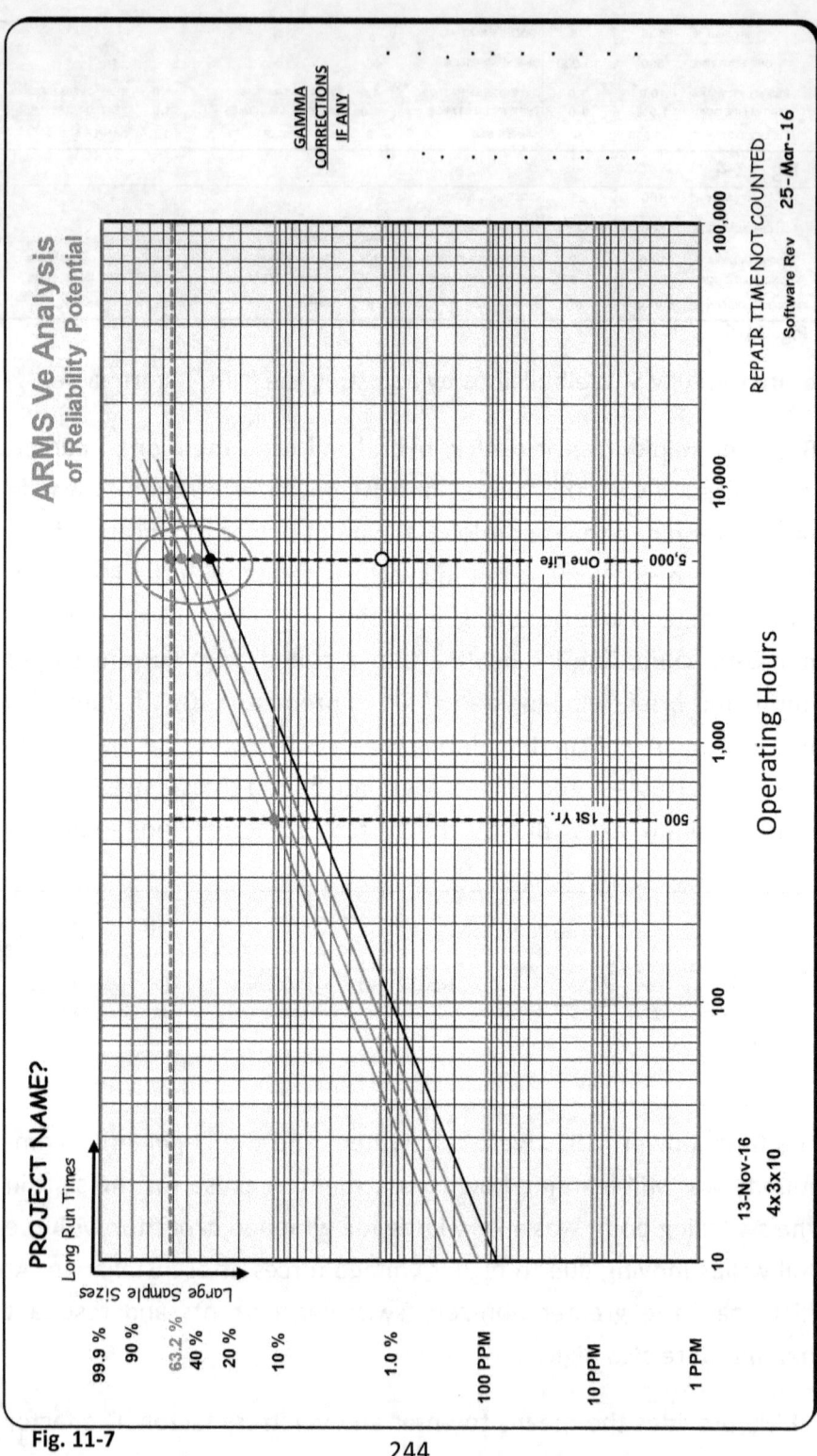

Fig. 11-7

244

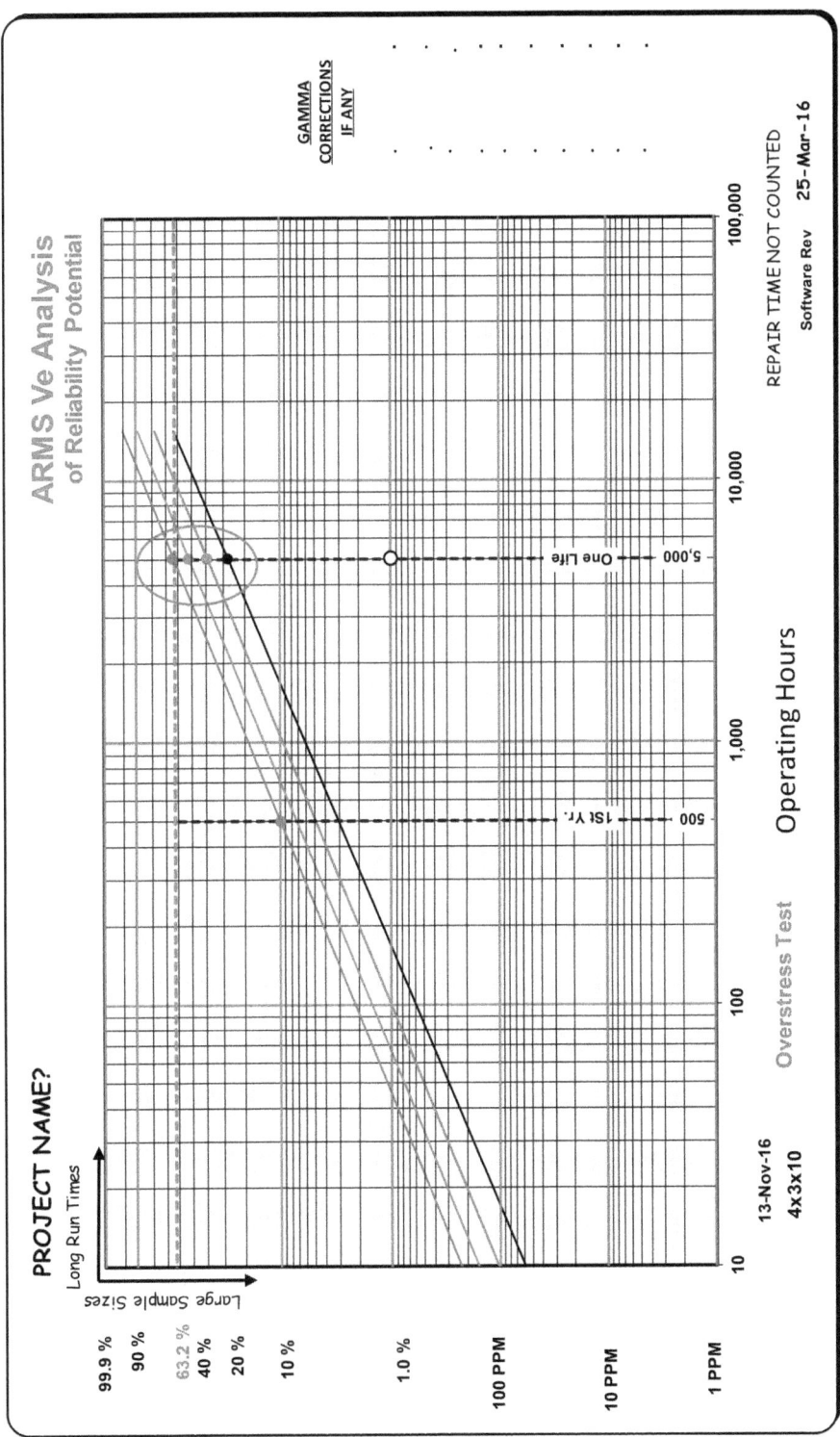

Fig. 11-8

245

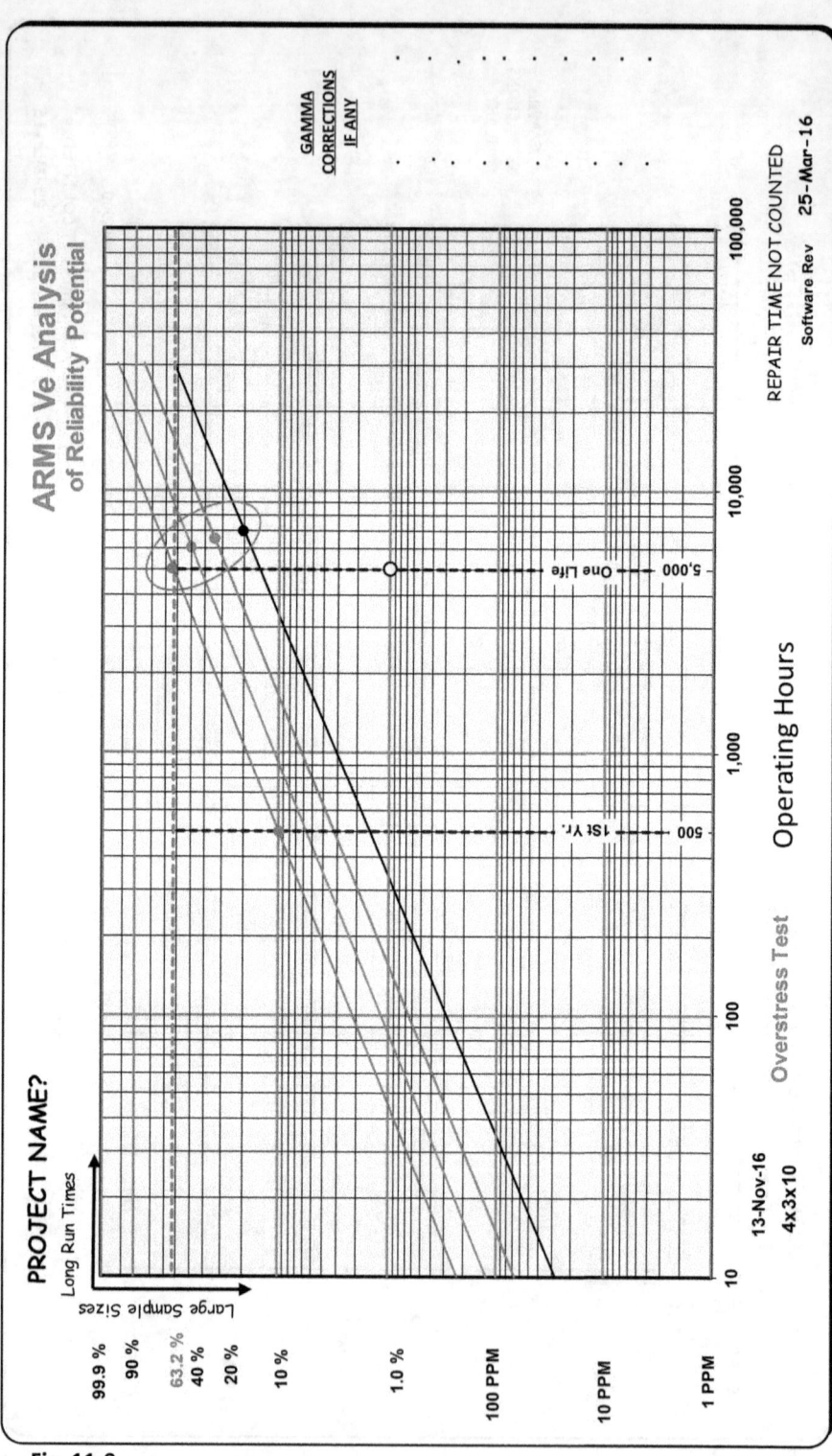

Fig. 11-9

246

is based on the increase in anticipated failures from an applied overstress. This can only be determined on a case by case basis. If there will not be overstress testing, enter a 1 into the overstress factor box **(Fig.11-4, 7).**

The INPUT sheet for this example shows overstress factors of 1.2, 1.3 & 1.4 have been entered in to the ARMS' model **(Fig.11-6,8).** The result of these entries shifts the end points of the test to the right by a corresponding factor. The shifted DOTS can be seen in the ARMS' Weibull graph within the red circles **(Fig. 11-7,8).** Over-stress testing results in enhanced α Alpha learning and a spread in the plotted ARMS' Weibull lines. Be conservative in the application of this overstress ARMS' feature. This is another judgement call, in our ARMS' "Fermi model."

Overstress testing brings to mind the benefits of finite element analysis. If available, FEA should be conducted on all design elements that are suspected of having a built-in overstress conditions. Planned overstress tests must avoid exceeding the stress capabilities of component parts; mechanical or electrical.

One case of built-in overstress that I was asked to evaluate concerned a series of plastic potentiometers used on the KH11 National Reconnaissance Office optical surveillance program, a defense critical program. This part was later replaced by electronics. At the time I was called in to help, the potentiometers were delaying the entire KH11 program. All attempts to solve the problem by the prime contractor, supplier and NRO personnel had failed. This situation was considered a serious national security problem. The potentiometers were electrically noisy and had a short life at low vacuum conditions in space. Surveillance rocket launches were being delayed.

The construction of the potentiometers involved silk screening a resistive carbon pattern on to the surface of a plastic slug that was at a "B" stage of curing. The silk screen pattern was designed to yield a specific electrical output when the potentiometer was rotated, a Sine, Cosine, Trapezoid, or Square wave output for example. After silk screening, the slug was heated under pressure to force the carbon "resist" into the surface of the plastic slug and to cure the plastic material, making it hard. This part of the process appeared to be in order, as surface hardness numbers were according to specification. A signal pick-up finger was formed from a single gold wire coming off of a solid arm mounted to the potentiometers rotary shaft. The design required manufacturing to adjust the gold wire form to achieve a specified force against the plastic material. Excessive wiper force on the surface would result in abrasive wear and a too light force would result in excessive electrical noise and arcing wear. The window of success was extremely tight. Manufacturing was taking 30 minutes to an hour to pass a single part. Production was, at times, not progressing and parts retested by the customer were failing to pass incoming tests at a high rate. The design and process was not workable and jeopardizing the KH11 mission. My prior experience with synchro's and resolvers for the aircraft instrument industry indicated that

Rate of deflection was calculated using a modified form of Castigliano's equation for cantilevered beams to compensate   for the curvature. $M_o = (3/2)PR \sin \theta$

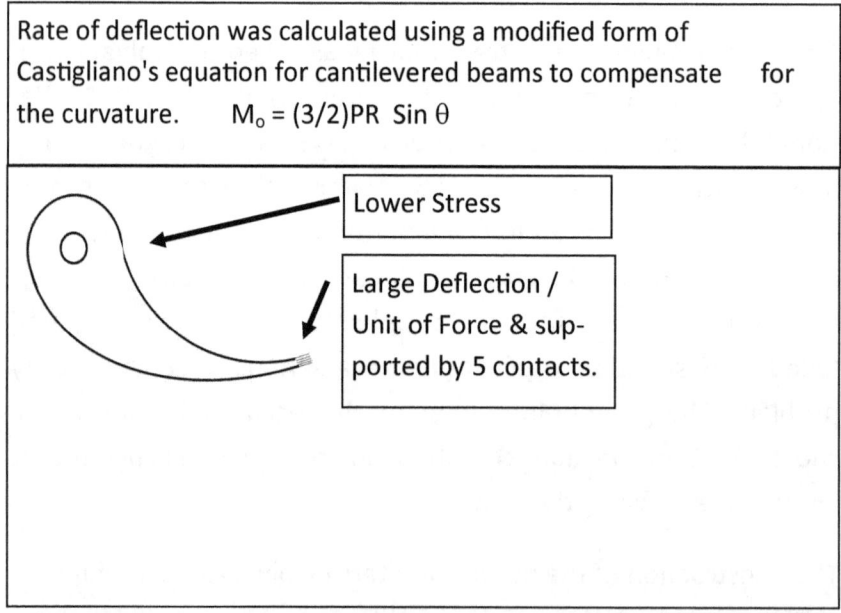

Lower Stress

Large Deflection / Unit of Force & supported by 5 contacts.

Fig. 11-10

the stress level being placed on the gold wire was exceeding gold's stress yield limit and the wire was yielding over time. This was easily confirmed by stress calculations of Young's modulus for the wire. Any introductory strength of materials textbook has the equations necessary to determine what was happening. The gold wiper arm material was overstressed and entering into "plastic yielding" which has time, heat and vibration elements. The wiper wire would not hold the applied stress and was relaxing in production and in route to the customer. There was no obvious solution. A heavier gauge gold wire would be too stiff causing an even smaller zone of adjustment and excessive forces. When existing potentiometers were overstress tested, using vibration and heat cycles, failures were immediate and a duplicate of conditions plaguing the customer. The solution to the problem was to separate some of the functions assigned to the gold wire. A new wiper form was made from thin sheet stainless steel in the form of a paisley leaf. The wide end was attached to the rotor and then tapered to the thin end where five small gold wires were attached, which provided contact to the plastic slug. Electrical noise in sliding contacts is reduced by $n/\text{SQRT}(n)$ where (n) is the number of electrical contacts. Ragnar Holm "Electric Contacts Theory and Application." With this configuration, the tip has a relatively large deflection range within the range of acceptable pressures delivered to the plastic slug. The wider range provides a greater latitude of adjustment during assembly. The paisley arm was pre-formed on a cone to a visually observed deflection and the potentiometer was assembled without the need for a force adjustment. A blind assembly was made directly visible before assembly. Sampling of the pressures obtained directly in production indicated **4σ** conformance to the specification as assembled. Assembly was reduced from nearly an hour to under a minute. The use of five independent contact wires reduced the electrical noise and electrical wear to levels never achieved on the program before.

Overstress durability testing did not result in failures during the specification life period. The new configuration was implemented and the customer supplied with the critical hardware just in time.

Overstress testing is a good way of uncovering weak points in a design as long as the failures revealed do not depart widely from reality.

Several examples of built in stress and stress risers were identified and designed out of the Sub-Roc Missile Arming Timers mentioned earlier. When the "Normal" environment includes 7000 G's of shock, you are working in a new reality and need to think out of the box. Using overstress as a tool to accelerate durability or to find weak points that are near the limits of the design is a special discipline requiring a wide range of experience and supported by a mathematical base. There are no simple answers, because each design is unique. Successful designs can carry over, but often do not.

Overstress testing has more benefits than accelerating failures and advancing the development time clock. There are often usage conditions never planned for in the design specifications that can occur and be survived if they had been a part of your overstress program. A case in mind occurred at the Bell Helicopter factory in Euless, Texas, during the latter part of the Vietnam War. I was attending a supplier meeting on reliability. As the engineer representing the Huey Cobra's Horizontal Situation Indicator (HSI), the Cobra II primary flight instrument, I was selected to take a flight in a Cobra II. But it was not a simple flight, where you receive a briefing and stroll out to the aircraft. The group of suppliers were to take a tour of the assembly plant, during which I would be taken to the flight line in a simulated war situation. The tour was fantastic, after reviewing the great detail of safety steps involved in the rotor assembly operation, where a section cut off from every rotor blade was put through a rip apart test, and the results were documented with the blade serial number. The blade was a laminated assembly, which also had about a six-inch brass leading edge. As they planned, a Klaxon horn went off BANAH-! BANAH-! BANAH-! and someone grabbed me from behind and shoved me through a side door of the rotor test area. A Huey Cobra II was right there next to the building. I turned and saw the guy shoving me was Bell's Chief Pilot who recently returned from Vietnam. I think he said his first name was Dwayne, but I could be wrong on that. Dwayne helped me up into the cockpit with a strong push in the "Butt," as a flight crew member tossed in a helmet. "The 'Cong' are coming!" he yelled out, "buckle up and get the helmet on, we are leaving!" It was a little hard to tell, but it seemed like the rotors were at speed on battery before I heard the jet engines kick in. We were off in a flash! The point being made was when you are under attack, don't look for a warm-up time for the instruments or anything else. This was before today's digital electronic instruments.

Dwayne climbed to about 1000 feet and loitered around waiting for three Jet Rangers to catch up with the other suppliers. Next, we headed down to a "chopper obstacle course," which was a double row of concrete steel reinforced poles set so a helicopter could let its body hang down between the poles with rotors above as the chopper weaved through the course. Not Dwayne, he dropped down so the rotors were below the tops of the poles and inches from the poles as he made his way weaving left and right through the course. He yelled something about maintaining total control over the headset. We proceeded to climb a few thousand feet at the max climb rate for the Cobra II. I was glad I had a light breakfast. Dwayne yelled all kinds of things in my headset, not many made sense at the time. We were hovering, then he said "Do you see them down there?" Me not being very smart, I said "Where?" "THERE!" And the Cobra II tipped nose down about 80 degrees and we went into a dive. Dwayne pulled out and climbed back up going into a full "loop-the-loop" and at the top of the maneuver, we were totally upside down. Body up, rotors down, and he continued to pull through the tight circle. I did not know a helicopter could do that! Dwayne said they don't tell the military pilots they can, the stresses are too high. Now that is a true overstress test. We dove toward a group of trees lining a small creek. Dwayne said they are in the trees! As we slipped under some trees and were moving along the stream bed. "DO YOU SEE THEM!" "No, where?" "THERE!" He rotated the Cobra II to the right, pointing to the bank. Rapidly, "ACK! ACK! ACK! ACK! ACK!" he yelled, then swung around and proceeded up stream. The trees were closing in as I noticed some green things swirling around. We were trimming the lower tree branches with the rotors! Thank GOD they have a leading brass edge. Eventually there was a clear spot overhead and we lifted out at max climb. After this, the trip back to the hanger seemed a little mundane! Later Dwayne explained how many of the elements of our flight were, in fact, at the worst-case stress conditions of what the Cobra II had to meet. This flight brought performance requirements in the specifications home to me to a degree I will never forget. The requirements are there for a reason and must be met!

Chapter Twelve

Additional ARMS Weibull
plotting examples:

- The survivors are expensive hardware that have received the benefit of a capital expense.

- The Failure Analyst should first examine the test units but the failure analysis needs to include a review of assembly processes, driven by the unit failure analysis clues.

- Facilitating Outstanding Leadership / Teamwork between all departments

- Engineering Design & Innovation Under Control

ADV BRAKES is an advanced braking system for high performance vehicles requiring greater braking force under extreme operating profiles, that generate significantly higher heat in the brake disks. The brake design had dual disks on each wheel. Grand Prix La Mons style racing and speed runs down Pike's Peak were brake profiles studied in setting up the durability test profile.

12 brake assemblies were tested until 4 failed as shown in the Validation Test table (FIG. 12-2). All were removed from test for detailed analysis. The ADV BRAKE test plan includes normal maintenance routines on brake pads, which could be changed as deemed necessary.   At this point, brake pad development was not a part of the ADV Brake design durability evaluation. All brake assemblies were submitted for Failure Analysis and the information collected was utilized by Simon in engineering to make mid-design generation improvements. The ADV BRAKE design and Failure Analysis details are confidential and are not available for presentation.

ARMS indicates a validation test Median Rank of 36.4%. This test data may be applied to your ARMS' model to obtain the same results and validate your copy of ARMS' software (Fig.12-4).

 The ADV BRAKE team's ARMS' model was adjusted to have the lowest data point fall on the (GEN2) Weibull line as shown (Fig.12-3). A beta vector of 4.0 was added to indicate a projection of the median characteristic wearout conditions, an indication of the expected β slope during failures from a larger sample size. As with the design samples tested, if additional samples from the same group were tested, they would be expected to start wearing out when the test duration intersects the plotted 4.0 slope. This gives us an idea of what to expect at higher or lower sample sizes. If applied with discretion, this is a powerful concept in that much higher sample sizes, which would be expensive to build and

PROJECT NAME?   10-Mar-17

IPT Start	12	PROCESS % SIGMA	33	
ONE YEAR TIME	500 (HOURS)	ONE YEAR KILOMETERS	25,000	
ONE LIFE in YEARS	10.0 (Kilometers)	OVERSTRESS FACTOR	1.0	TEST OVERSTRESS 1.1 / 1.2
ONE LIFE GOAL	5.0%	TEST IN LIFETIMES	1.1 / 1.2 / 1.3	
ONE LIFE DISTANCE	250,000	SAMPLE SIZE	12 / 12 / 12	

Program Manager – Name & Workshop DATE
ARMS Ve Engineer – Name & Tel #?

ADV BRAKE SYSTEM

Next ARMS Review date?   Workshop Date >>> 01-Jan-00

Item	D.O.E. / TRIZ	FEA / GOLDFIRE INNOV.	RED X / ENG. STUDY	GEN 1 Alpha α	Support Factor	GEN 1 IPT % W/O (SF)	GEN 1 IPT % W (-SF)	GEN 2 Alpha α	Support Factor	GEN 2 IPT % W/O (SF)	GEN 2 IPT % W (-SF)	GEN 3 Alpha α	GEN 3 IPT % W/O (SF)	GEN 3 IPT % W (-SF)
KNUCKLE	5%			1.000		0.45	0.47	1.091		0.35	0.35	1.091	0.245	0.245
	100%													
CASTING		10%				0.04	0.05			0.03	0.03		0.020	0.020
HYDRAULICS		80%				0.36	0.38			0.28	0.28		0.201	0.201
BEARING ASM		10%				0.05	0.05			0.03	0.03		0.024	0.024
CALIPER	32.5%					2.74	2.95			1.98	1.98		1.272	1.272
	100%													
BRIDGE ASM		35%				0.95	1.06			0.69	0.69		0.435	0.435
PAD ASM		60%				1.66	1.76			1.18	1.18		0.759	0.759
BRIDGE BOLTS		5%				0.13	0.13			0.10	0.10		0.078	0.078
DISC ASM	32.5%					2.54	2.69			1.97	1.97		1.372	1.372
	100%													
DISC		80%				2.03	2.13			1.54	1.54		1.060	1.060
SPRINGS		20%				0.51	0.56			0.43	0.43		0.312	0.312
HUB ASM	30%					2.43	2.68			1.70	1.70		1.019	1.019
	100%													
HUB		95%				2.29	2.54			1.60	1.60		0.943	0.943
ABS RING		1%				0.03	0.03			0.03	0.03		0.022	0.022
WHEEL STUDS		4%				0.11	0.11			0.08	0.08		0.054	0.054
12 ms @ One Year IPT							8.2				6.0			3.91

Fig. 12-1

Line #	QTY TAKEN OFF TEST	Failure=1 Suspension=0	Rank Individual TIMES or CYCLES at Test Termination — MILES	INCREMENT (< Acceleration Factor)	Mean Order Number	Previous Mean Order Number (Beta Est_1 >>)	Median Rank %	Plot Point Y	Plot Point X	Sample Size 95% Conf (+)	Sample Size 95% Conf (-)
TEST 1	TOTAL 12	1Γ = 1	1		Program Name	1.09	4.00	1	178,521	"Applied Life Data Analysis" Wayne Nelson +θ	1 "Applied Life Data Analysis" Wayne Nelson −θ
1	1	1	105,589	1.000000	1.000		5.613%	-2.8514	105,589		65,673
2	1		120,304	1.090909		10.00		-2.8514	105,589		65,673
3	1	1	131,245	1.000000	2.091		14.417%	-1.8600	131,245		81,630
4	1		133,531	1.212121		8.00		-1.8600	131,245		81,630
5	1		141,149	1.363636		7.00		-1.8600	131,245		81,630
6	1	1	157,096	1.000000	3.455		25.422%	-1.2265	157,096		97,708
7	1	1	178,521	1.000000	4.818		36.427%	-0.7919	178,521		111,034
8	5		250,000					-0.7919	178,521		111,034
9											
10											
11											
12											

Fig. 12-2

256

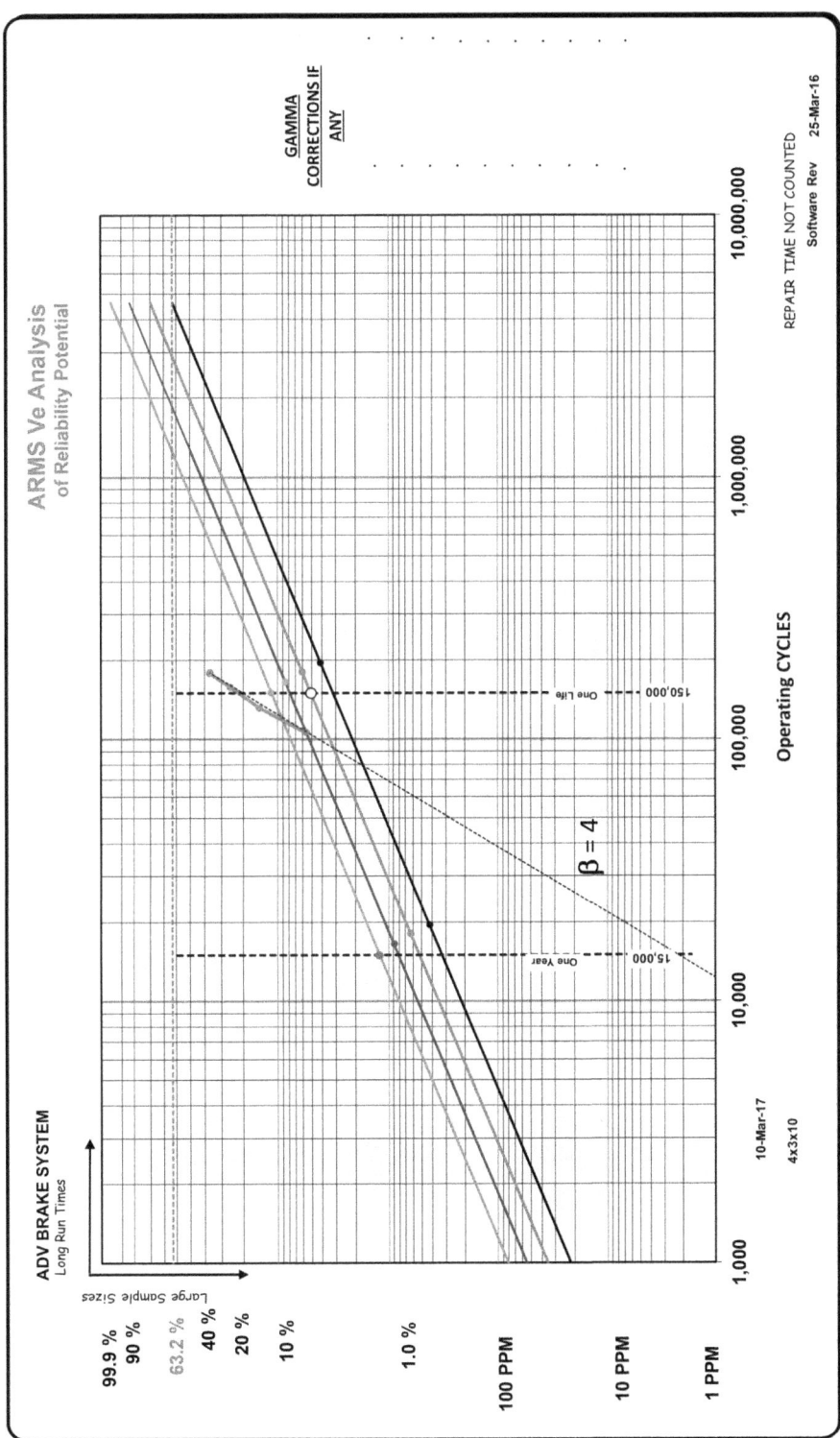

Fig. 12-3

257

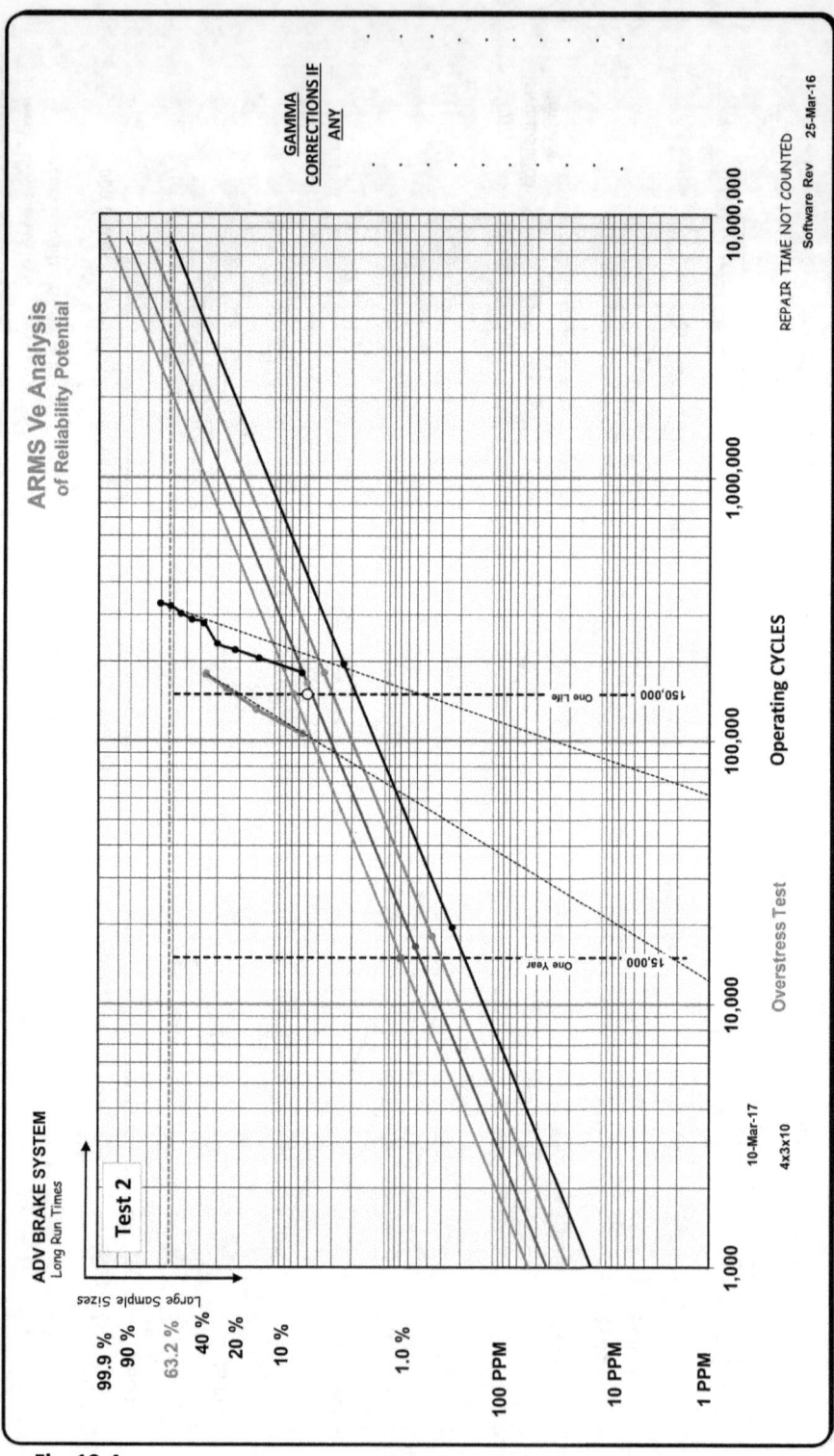

Fig. 12-4

258

test, would have results that are not expected to be any better than shown here with projections. In a conservative approach, only a lower 1.645σ median limit (to the left of the median) should be placed on the failure data. A larger sample size would tighten the limit spread, but it is advisable to make design improvements based on the failure analysis and test a new design with a similar small sample size of 12 at a lower cost. Progress comes from rapid design iterations with reasonably small sample sizes.

After reviewing the nature of the failures and the condition of the survivors, it was decided not to modify the ARMS' model yet. Additional data analysis with an ARMS' Weibull was conducted making use of the failure analysis results, which revealed two clear modes of failure. The Validation Test data was split into two

TEST 2A		3 Γ =			Program Name				
TOTAL 12						Beta Est_3 >>	8.50	1	233,043
Line #	QTY TAKEN OFF TEST	Failure=1 Suspension=0	Rank Individual Measures at Test Termination	INCREMENT	Mean Order Number	Previous Mean Order Number	Median Rank %	Plot Point Y	Plot Point X
			MILES		1.00				
1	1	1	180640	1.000000	1.000		5.613%	-2.8514	180,640
2	1	1	205323	1.000000	2.000		13.683%	-1.9163	205,323
3	1	1	220745	1.000000	3.000		21.753%	-1.4053	220,745
4	1	1	233043	1.000000	4.000		29.824%	-1.0380	233,043
5	8							-1.0380	233,043

Fig. 12-5

TEST 2B		4 Γ =			Program Name				
TOTAL 8						Beta Est_4 >>	11.00	1	330,820
Line #	QTY TAKEN OFF TEST	Failure=1 Suspension=0	Rank Individual TIMES or CYCLES at Test Termination	INCREMENT	Mean Order Number	Previous Mean Order Number	Median Rank %	Plot Point Y	Plot Point X
			MILES		1.00				
1	1	1	279195	1.000000	1.000		8.300%	-2.4460	279,195
2	1	1	288205	1.000000	2.000		20.214%	-1.4880	288,205
3	1	1	303510	1.000000	3.000		32.128%	-0.9479	303,510
4	1	1	324110	1.000000	4.000		44.043%	-0.5437	324,110
5	1	1	330820	1.000000	5.000		55.957%	-0.1984	330,820
6	3							-0.1984	330,820

Fig. 12-6

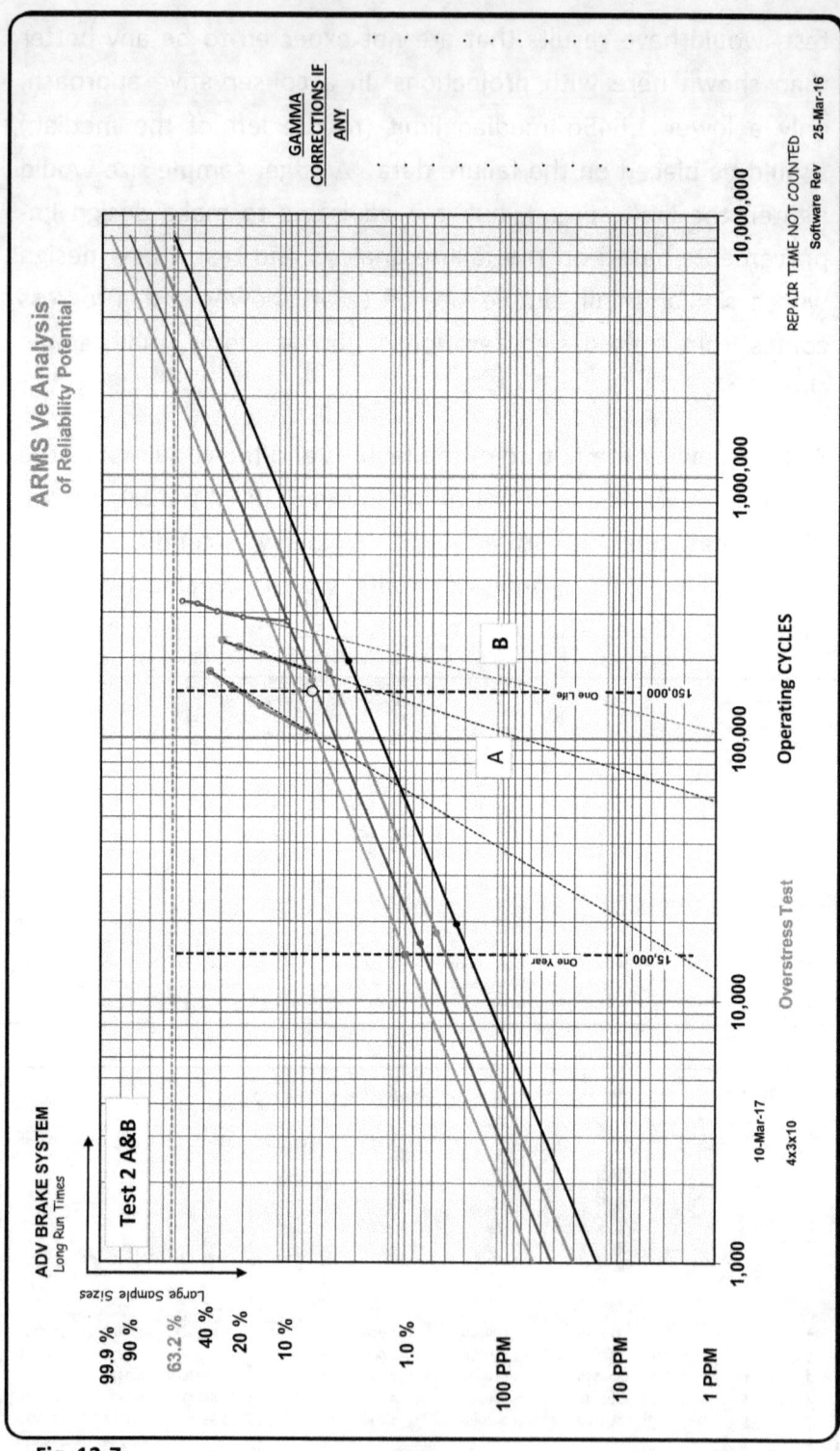

Fig. 12-7

260

groups **(Fig. 12-5, 12-6)** and graphed. There were 4 failures at an earlier number of cycles. These 4 failures plus the 8 that survived testing up to that point are treated as a set.

This set is noted as (Test 2A). The remaining 5 failures and 3 survivors become a set (Test 2B). Both failure mode median failures are to the right of the interim end of life 5% goal **(Fig. 12-7)**.

During the next design cycle, adjustments necessary to satisfy production requirements and a more demanding goal were made. Normally, actions of this type result in a set back of the test results and require additional design action. In this case, when an additional 12 brakes were tested after production update by the Manufacturing Engineering Group, results were very favorable. Detailed failure analysis resulted in a few additional improvements not reported here. Additional cycles were completed to confirm all design and manufacturing adjustments being made to lower costs.

TEST 1		∫Γ=		Program Name					Sample Size 95% Conf (+)	Sample Size 95% Conf (-)
TOTAL 6					Beta Est.1 >> 1.40	1	572		1	1

Line #	QTY TAKEN OFF TEST	Failure=1 Suspension=0	Rank Individual TIMES or CYCLES at Test Termination	INCREMENT	Mean Order Number	NIBPSS	Median Rank %	Plot Point Y	Plot Point X	"Applied Life Data Analysis" Wayne Nelson	"Applied Life Data Analysis" Wayne Nelson
			HOURS		1.20					+θ	−θ
1	1	1	112	1.000000	1.000		10.910%	−2.1583	112	219	57
2	1		213	1.200000		4.00		−2.1583	112	219	57
3	1	1	250	1.000000	2.200		29.673%	−1.0441	250	489	128
4	1		484	1.600000		2.00		−1.0441	250	489	128
5	1		500	2.400000		1.00		−1.0441	250	489	128
6	1	1	572	1.000000	4.600		67.200%	0.1086	572	1,120	292
7											
8											

Fig. 12-8

Most text books calculate Median Rank with Benard's Approximation which provides a sufficiently accurate result for most work: MR = (MON-0.3)/(N+0.4)

ARMS employs a slightly more accurate series formula:

$$MR = 1 - 2^{\left(\frac{1}{N}\right)} + \left(\left(\frac{MON-1}{N-1}\right) * \left(2^{\left(1-\frac{1}{N}\right)}-1\right)\right)$$

Originally, the scale of the horizontal axis was 10—100,000 and our plot was off center. The ARMS' Weibull graph was adjusted to 10—10,000 hours by highlighting the axis and right clicking to bring up an axis format screen. The horizontal axis can be set to a range of base ten numbers. Notice the Weibull graph has three cycles and earlier the ARMS' Weibull graph had four cycles.

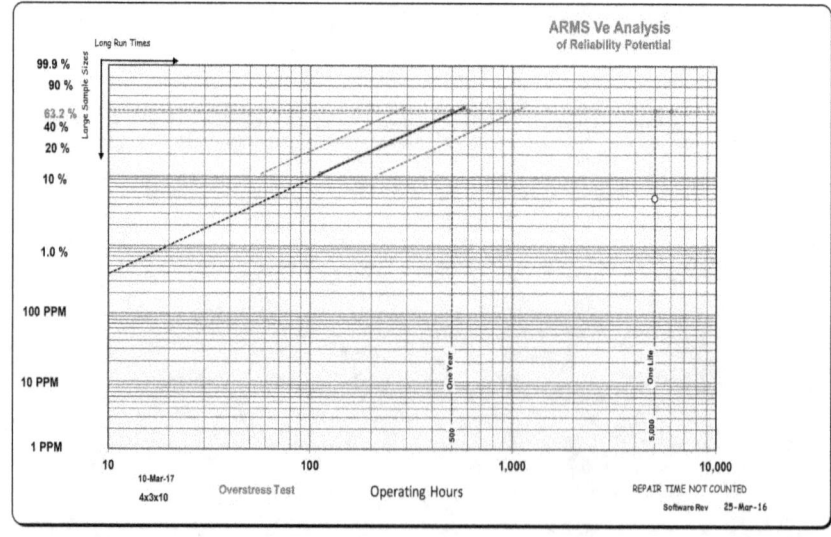

Fig. 12-9

7 units tested with 3 suspensions followed by 4 failures were placed into ARMS and yielded MR 82.458% **(Fig.12-10)**. The β slope of the plotted data is manually estimated with the Beta est.>> input of 1.25. The Weibull plot of the data **(Fig.12-9)** indicates there may be problems with how the units were put into service and how the data was collected. We know nothing about the condi-

TEST 1		IΓ=				Program Name					Sample Size 95% Conf (+)	Sample Size 95% Conf (-)
TOTAL	7					Beta Est.1 >>	1.25	1	880		1	1
Line #	QTY TAKEN OFF TEST	Failure=1	Suspension=0	Rank Individual TIMES or CYCLES at Test Termination	INCREMENT	Mean Order Number	NIBPSS	Median Rank %	Plot Point Y	Plot Point X	"Applied Life Data Analysis" Wayne Nelson	"Applied Life Data Analysis" Wayne Nelson
				HOURS		1.33					+ θ	− θ
1	1			100	0.952381		6.00					
2	1			110	1.333333		5.00					
3	1			130	1.600000		4.00					
4	1	1		150	1.000000	1.600		17.542%	-1.6457	150	279	81
5	1	1		320	1.000000	3.200		39.181%	-0.6986	320	596	172
6	1	1		540	1.000000	4.800		60.819%	-0.0651	540	1,006	290
7	1	1		880	1.000000	6.400		82.458%	0.5542	880	1,639	473
8												

Fig. 12-10

tions leading up to the data set, which is not a good way to start an analysis. The pattern of failures is of the type often seen in field service applications. Actual hours of operation before the first failure is suspect. The three reported as taken off early without failure may in fact be sales return product. The next unit failing at 150 hours may actually be a unit installed but not turned on, then failing within a few hours. This is typical of what happens with HVAC field units. They get reported as on when placed in position but power is not applied until later. I suspect the data. The plotted pattern is an apparent bi-modal result. The previous discussion involved a bi-modal application, however, this may be due to bi-model failure conditions within a component, system design, or physical applications. Failure analysis, extended to application conditions, is required to ascertain all causes. Tools like the 3D Warranty graph can be useful when filtered by system designs, GPS data, or distribution and Installing suppliers.

For a set of 131,494 systems **(Fig.12-11,12)**, ARMS yields a maximum Median Rank of 6.341%. The β slope of the plotted data is manually estimated with the Beta est.>> input of 1.80.

A large sample size improves MR calculations and the 95% range of the Median very closely follows the Median line.

	TEST 1		I Γ =			Program Name				
TOTAL	131,494						Beta Est.1 >>	1.80	1	12,000
Line #	QTY TAKEN OFF TEST	Failure=1 Suspension=0	Rank Individual TIMES or CYCLES at Test Termination HOURS	INCREMENT	Mean Order Number 481.00	NIBPSS	Median Rank %	Plot Point Y	Plot Point X
1	216		1,000	0.997981		131278.00			
2	481	1	1,000	1.000000	480.029		0.365%	-5.6117	1,000
3	969		2,000	1.009135		129828.00		-5.6117	1,000
4	840	1	2,000	1.000000	1327.702		1.009%	-4.5907	2,000
5	3160		3,000	1.034478		125828.00		-4.5907	2,000
6	453	1	3,000	1.000000	1796.321		1.366%	-4.2865	3,000
7	5820		4,000	1.084836		119555.00		-4.2865	3,000
8	419	1	4,000	1.000000	2250.867		1.712%	-4.0592	4,000
9	8163		5000	1.164634		110973.00		-4.0592	4,000
10	380	1	5,000	1.000000	2693.428		2.048%	-3.8779	5,000
11	10265		6,000	1.283792		100328.00		-3.8779	5,000
12	377	1	6,000	1.000000	3177.418		2.416%	-3.7108	6,000
13	11516		7,000	1.450965		88435.00		-3.7108	6,000
14	409	1	7,000	1.000000	3770.863		2.867%	-3.5372	7,000
15	11884		8,000	1.677425		76142.00		-3.5372	7,000
16	336	1	8,000	1.000000	4334.477		3.296%	-3.3957	8,000
17	11597		9,000	1.980385		64209.00		-3.3957	8,000
18	346	1	9,000	1.000000	5019.691		3.817%	-3.2463	9,000
19	10765		10,000	2.381877		53098.00		-3.2463	9,000
20	332	1	10,000	1.000000	5810.474		4.419%	-3.0968	10,000
21	9731		11,000	2.920451		43035.00		-3.0968	10,000
22	330	1	11,000	1.000000	6774.223		5.151%	-2.9396	11,000
23	8654		12,000	3.662656		34051.00		-2.9396	11,000
24	427	1	12,000	1.000000	8338.177		6.341%	-2.7256	12,000
25	33624		13,000			33624.00		-2.7256	12,000

Fig. 12-11

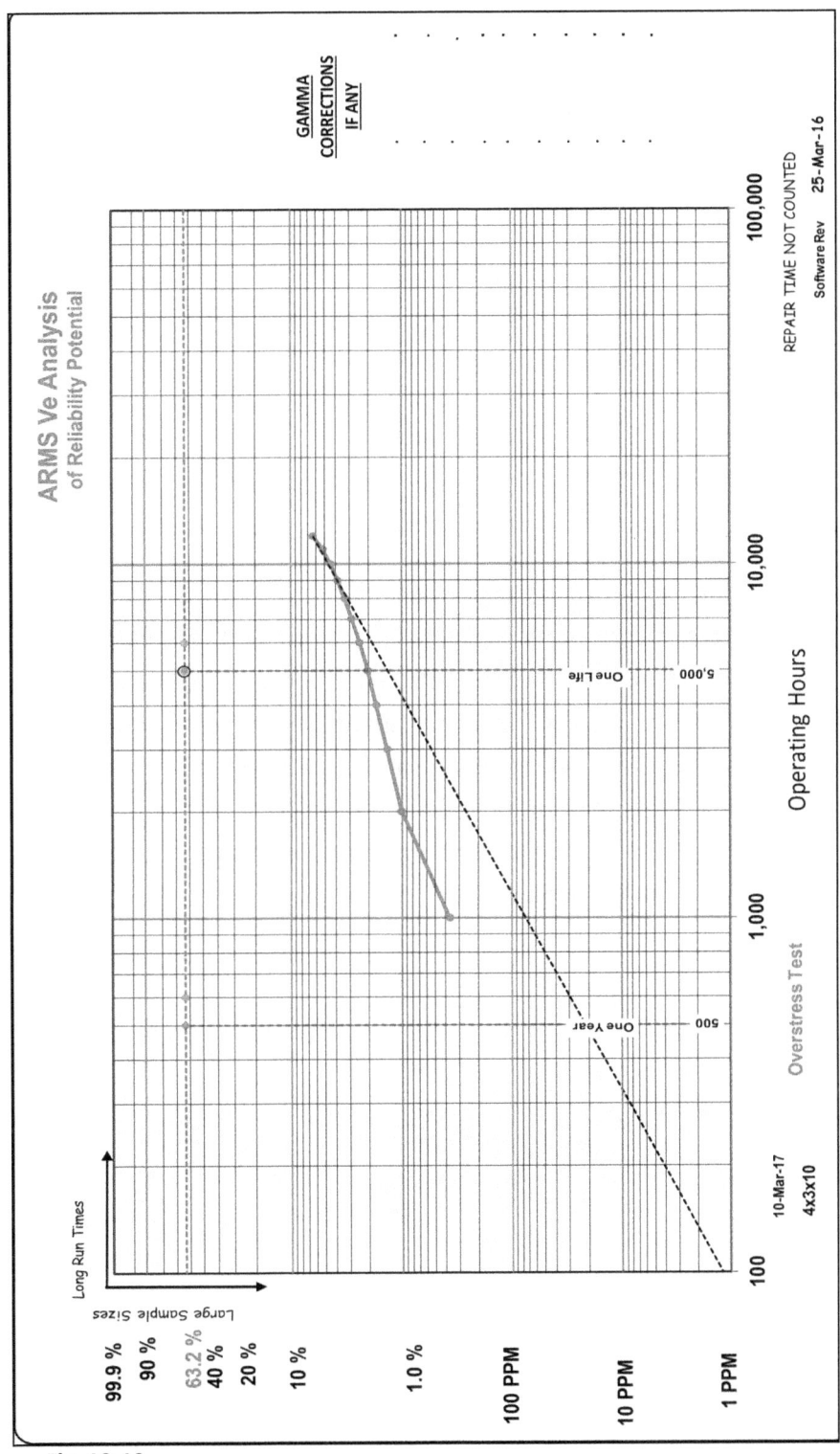

Fig. 12-12

For a set of 20 units **(Fig.12-14,15)**, ARMS yields a maximum median rank of 23.025%. The β slope of the plotted data is manually estimated with the Beta est.>> input of 1.5.

Line #	QTY TAKEN OFF TEST	Failure=1 Suspension=0	Rank Individual TIMES or CYCLES at Test Termination	INCREMENT	Mean Order Number	NIBPSS	Median Rank %	Plot Point Y	Plot Point X
TEST 1	**TOTAL** 20		1 Γ =			Program Name	Beta Est.1 >> 1.50	1	1,197
			HOURS		1.00				
1	1	1	296	1.000000	1.000		3.406%	-3.3622	296
2	1	1	605	1.000000	2.000		8.311%	-2.4445	605
3	1	1	798	1.000000	3.000		13.216%	-1.9537	798
4	1	1	1,035	1.000000	4.000		18.120%	-1.6099	1,035
5	1	1	1,197	1.000000	5.000		23.025%	-1.3406	1,197
6	15							-1.3406	1,197
7									
8									

Fig. 12-13

We are looking at the bottom of a bath tub. The units should have been tested to failure with overstress and possibly extended testing to insure failures. The 15 units should be performance tested, one or two possibly sent to failure analysis as survivors which may exhibit impending failure and provide valuable information. All too often I have seen units like the 15 survivors put on the shelf. **The survivors are expensive hardware that have received the benefit of a capital expense.**

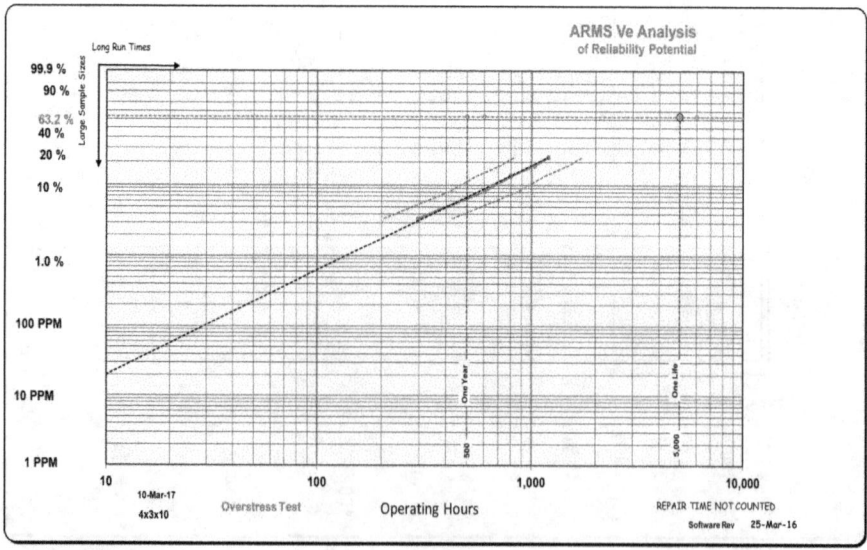

Fig. 12-14

Conditions in the units can yield information on imminent failure or over design that, if understood, could save a lot of engineering budget or product cost. Are the failures and survivors the same, or do the survivors have characteristics which allowed them to live?

For a set of 10 units (Fig.12-15,16), ARMS yields a maximum median rank of 45.90%. The β slope of the plotted data is manually estimated with the Beta est.>> input of 4.4.

If you want a challenge, pick up a pencil and calculate the Median Ranks for this data set.

Line #	QTY TAKEN OFF TEST	Failure=1 Suspension=0	Rank Individual TIMES or CYCLES at Test Termination	INCREMENT	Mean Order Number	NIBPSS	Median Rank %	Plot Point Y	Plot Point X	"Applied Life Data Analysis" Wayne Nelson +θ	"Applied Life Data Analysis" Wayne Nelson -θ
TEST 1		I Γ=			Program Name					Sample Size 95% Conf (+)	Sample Size 95% Conf (-)
TOTAL	10					Beta Est.1 >>	4.40	1	670	1	1
			HOURS		1.11						
1	1	1	400	1.000000	1.000		6.697%	-2.6691	400	673	238
2	1		450	1.111111		8.00		-2.6691	400	673	238
3	1	1	490	1.000000	2.111		17.389%	-1.6553	490	824	291
4	1		500	1.269841		6.00		-1.6553	490	824	291
5	1		530	1.481481		5.00		-1.6553	490	824	291
6	1	1	589	1.000000	3.593		31.845%	-0.9664	589	991	350
7	1	1	670	1.000000	5.074		45.901%	-0.4872	670	1,127	398
8	1		700	1.975309		2.00		-0.4872	670	1,127	398
9	1		730	2.962963		1.00		-0.4872	670	1,127	398
10	1		750					-0.4872	670	1,127	398
11											

Fig. 12-15

The β slope of 4.4 suggests this set of units is in an end of life wearout condition. That could typically be mechanical wear, oil film degradation, chemical depletion, thermal excess or a combination of issues. The data seems stable suggesting a single cause of failure which needs to be confirmed by failure analysis first.

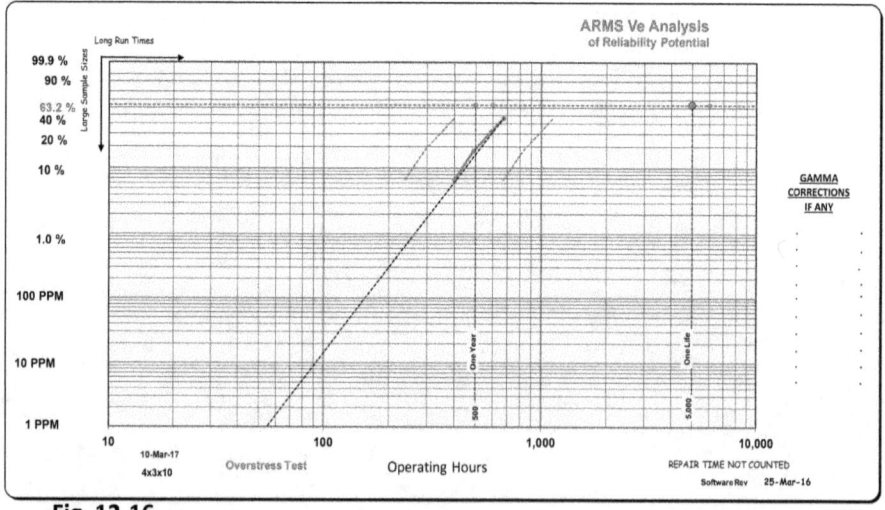

Fig. 12-16

For a set of 10 units (**Fig.12-17,18**), ARMS yields a maximum Median Rank of 93.303%. The β slope of the plotted data is manually estimated with the Beta est.>> input of 2.3.

TEST 1			Iⲅ=			Program Name					Sample Size 95% Conf (+)	Sample Size 95% Conf (-)
TOTAL	10					Beta Est.1 >>	2.30	1	1,400		1	1
Line #	QTY TAKEN OFF TEST	Failure=1 Suspension=0	Rank Individual TIMES or CYCLES at Test Termination	INCREMENT	Mean Order Number	NIBPSS	Median Rank %	Plot Point Y	Plot Point X		"Applied Life Data Analysis" Wayne Nelson	"Applied Life Data Analysis" Wayne Nelson
			HOURS		1.00						+ θ	– θ
1	1	1	200	1.000000	1.000		6.697%	-2.6691	200		336	119
2	1	1	370	1.000000	2.000		16.320%	-1.7250	370		622	220
3	1	1	500	1.000000	3.000		25.943%	-1.2029	500		841	297
4	1	1	620	1.000000	4.000		35.566%	-0.8221	620		1,043	369
5	1	1	730	1.000000	5.000		45.189%	-0.5087	730		1,228	434
6	1	1	840	1.000000	6.000		54.811%	-0.2303	840		1,413	499
7	1	1	950	1.000000	7.000		64.434%	0.0332	950		1,598	565
8	1	1	1,050	1.000000	8.000		74.057%	0.2996	1,050		1,766	624
9	1	1	1160	1.000000	9.000		83.680%	0.5949	1,160		1,952	690
10	1	1	1,400	1.000000	10.000		93.303%	0.9946	1,400		2,355	832
11												

Fig. 12-17

The earlier β slope appears to be about 1, indicating normal random failure followed by some wearout conditions. I would expect additional units put on to test will run into a "wall" around 2000 hours when wearout dominates the results. This is a bathtub result where you can see the bathtub forming. There may be a need for a Gamma adjustment. Adding a Gamma value of 150 hours causes the Weibull plot to form a straight line (**Fig.12-19**) with a β slope of 2.5. This can occur if the product has a condition consuming some

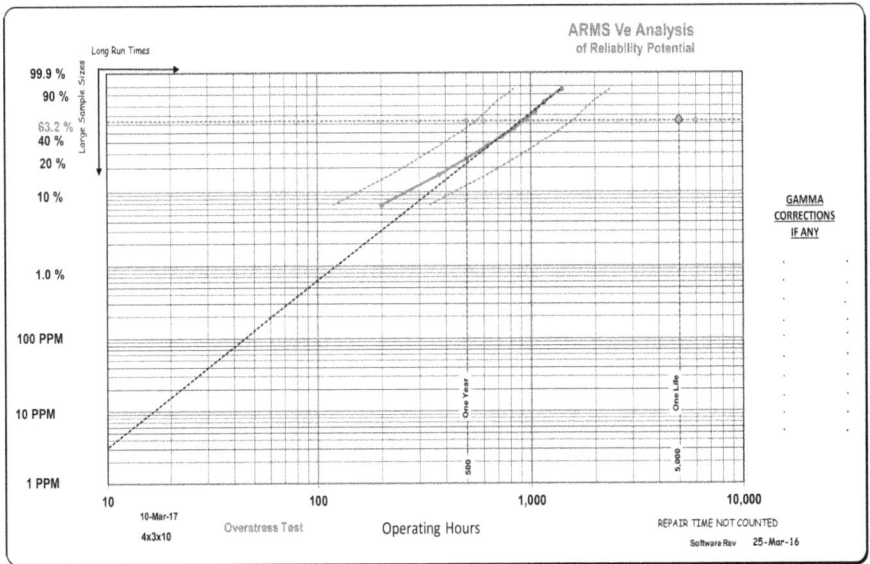

Fig.12-18

of the early life. The only way to know is to investigate how the data was collected and the treatment the product received prior to being placed into service or on test.

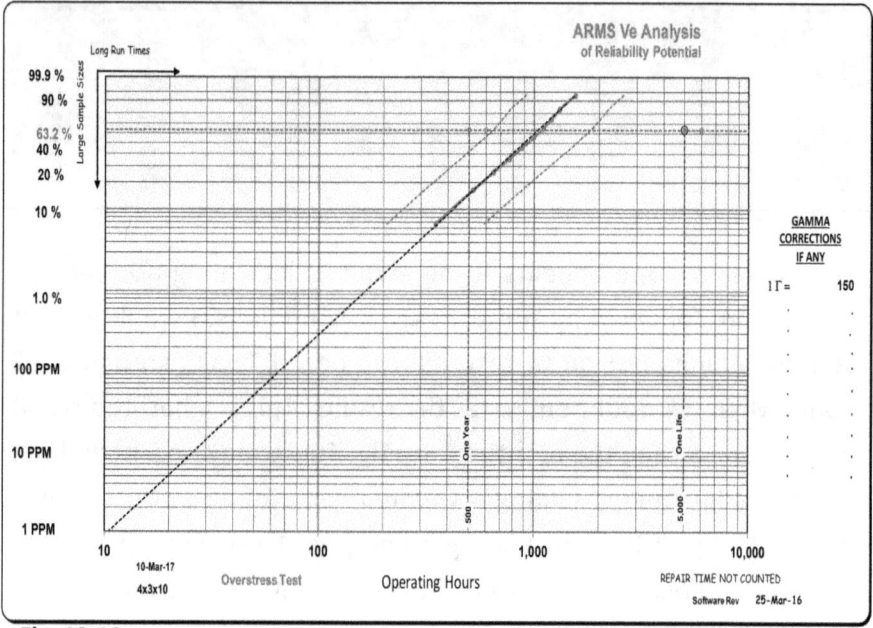

Fig. 12-19

For a set of 10 units **(Fig.12-20,21)** ARMS yields a maximum Median Rank of 54.811%. The β slope of the plotted data is manually estimated to have Beta = 1.0.

List suspensions before failures if times are equal.

Line #	QTY TAKEN OFF TEST	Failure=1	Suspension=0	Rank Individual TIMES or CYCLES at Test Termination	INCREMENT	Mean Order Number	NIBPSS	Median Rank %	Plot Point Y	Plot Point X	"Applied Life Data Analysis" Wayne Nelson +θ	"Applied Life Data Analysis" Wayne Nelson −θ
TEST 1	TOTAL 10		IΓ = 1.0	< Acceleration Factor	Line calc: 255, 0, 6	Beta Est.1 >> 1.00		PROJECT NAME?		120	Sample Size 95% Conf (+) 1	Sample Size 95% Conf (−) 1
				HOURS		1.00					+θ	−θ
1	1	1		10	1.0000	1.000		6.697%	-2.6691	10	17	6
2	1	1		26	1.0000	2.000		16.320%	-1.7250	26	44	15
3	1	1		42	1.0000	3.000		25.943%	-1.2029	42	71	25
4	1	1		65	1.0000	4.000		35.566%	-0.8221	65	109	39
5	1	1		90	1.0000	5.000		45.189%	-0.5087	90	151	53
6	1	1		120	1.0000	6.000		54.811%	-0.2303	120	202	71
7	4			120					-0.2303	120	202	71
8												
9												

Fig. 12-20

This would be the bottom of a shallow bath tub. True "End of Life" conditions would not have been encountered. In a real situation, failure analysis of the failed and surviving units taken off of test would be analyzed for important clues. **The Failure Analyst should first examine the test units but the failure analysis needs to include a review of assembly processes, driven by the unit failure analysis clues. If possible, after engineering design and process corrective actions, a larger sample tested to failure is recommended.**

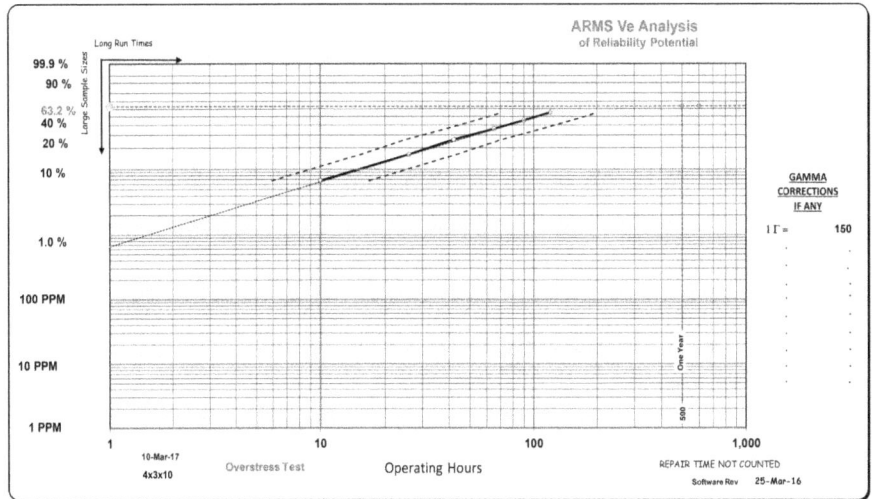

Fig. 12-21

271

Chapter Thirteen

ARMS Software Elements

- ARMS is a GUI interface supporting engineers in their efforts to apply statistics to design development efforts.

- Engineering visions are always ahead of hardware execution and validation tests.

- ARMS is the only program that I am aware of that will achieve this level of support to the technical management of an advanced development project.

- Facilitating Outstanding Leadership / Teamwork between all departments

- Engineering Design & Innovation Under Control

Recapping for Dr. Ebbinghaus:

ARMS (Allocated Risk Management System) was structured to reduce the pain and complexity of applying statistical methods to the process of reducing risks and improving the reliability of designs.

ARMS is a GUI interface supporting engineers in their efforts to apply statistics to design development efforts.

ARMS is primarily applied to advanced development engineering programs and programs applying new technology to existing designs. The applications range from complex systems with several sub-suppliers down to single device developments. Applications have included Hybrid Vehicles, Solid Oxide Fuel Cells (SOFC), Advanced Floating Disk Brake Systems, Electronic Steering Systems, Stair Climbing Wheel Chairs, Non-Plasma Reactors, Stirling Thermal Engines, Power Electronics Bays, Ceramic Sensors, Tire Air Pressure Sensors, Medical Electro-Mechanical Implants and a long list of automotive systems. Applications involved several hundred engineers working in team cells of from 20 down to one engineer. Most were teams of 6-10 engineers.

To introduce the topic of reliability improvement during advanced development was novel and new to all of the industries where ARMS was applied. The application of statistics to shape engineering thinking and planning had not been attempted, let alone achieved by other methods. ARMS is a unique approach that has been proven highly effective in anticipating and reducing high risks in design elements, when anomalies are less costly to correct. Changes are earlier, thereby reducing total design development time. To achieve this, ARMS had to gain acceptance with the engineering community and had to take the pain and confusion out of applying statistical methods as promised. ARMS recognizes that engineers start with the formation of visions and work to verify those visions through the application of mathematics and confirmation hardware in durability trials.

Engineering visions are always ahead of hardware execution and validation tests.

To have the most impact on the final outcome, ARMS works primarily in the region of Engineering Vision and design formation. To do so, ARMS pushes the required statistics into the background by creating a GUI (Graphical User Interface) with the engineers. This allows the Engineering Team to evaluate variations in design approach based on their combined knowledge before committing to

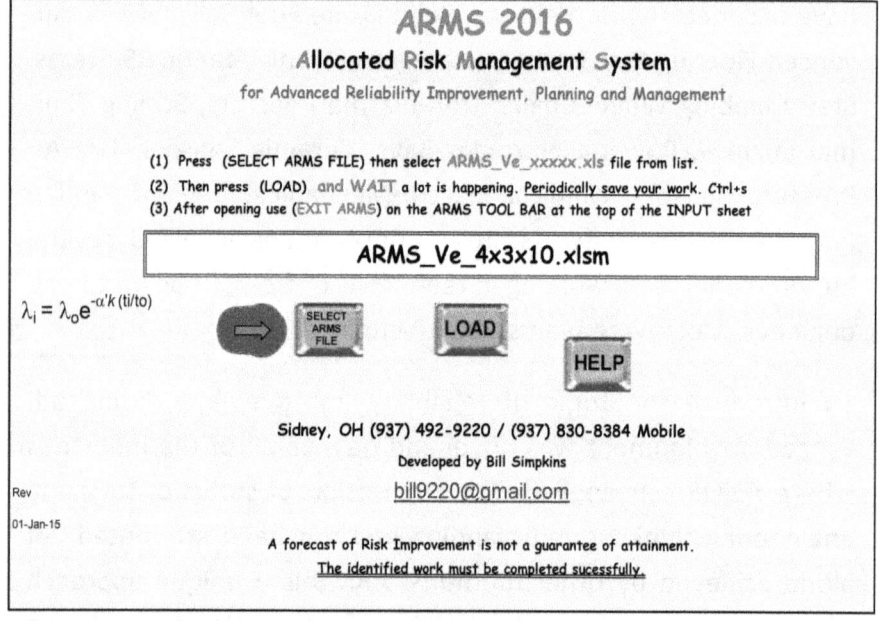

Fig. 13-1

budget expenditure. Decisions to advance or delay the development of design elements or when to buy capital equipment, services or structured studies like Design of Experiments can be evaluated with an ARMS' model.

ARMS is the only program that I am aware of that will achieve this level of support to the technical management of an advanced development project.

With these objectives in mind, an ARMS interface was developed.

Backup the following files which are the basic ARMS' software:

arms-ve-4x3x10,   arms-ve-5x4x6,

arms-ve-6x4x5,   arms-ve-6x5x8,

arms-ve-8x5x6.   load-arms-ve-2017

tables-ve-4x3x10,   tables-ve-5x4x6,

tables-ve-6x4x5,   tables-ve-6x5x8,

tables-ve-8x5x6.

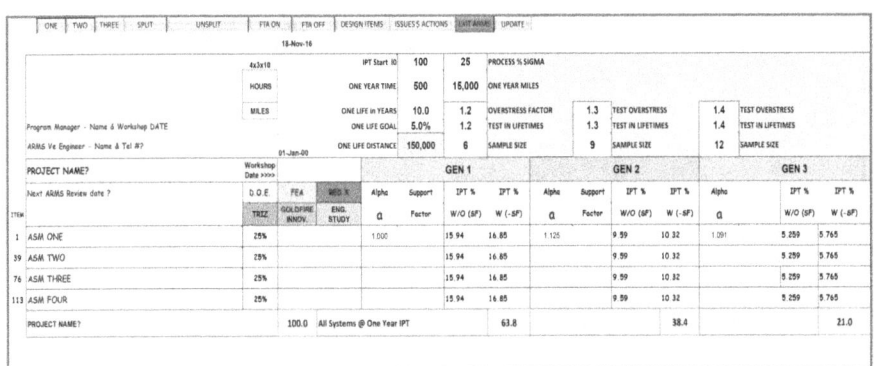

Fig. 13-2

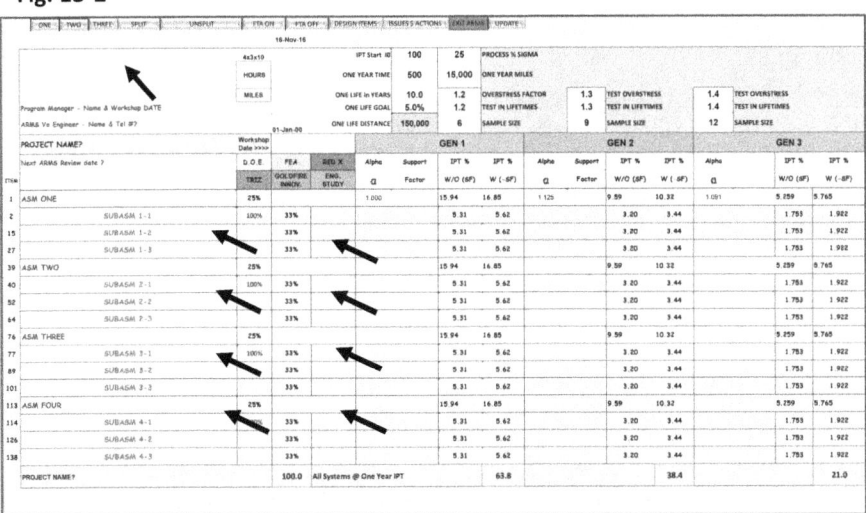

Fig. 13-3

Open **load-arms-ve- 2017 with** Excel. Press Select ARMS' File and search File Explorer for the matrix to be applied to your design. e.g. arms-ve-4x3x10. Press LOAD to load the files and open

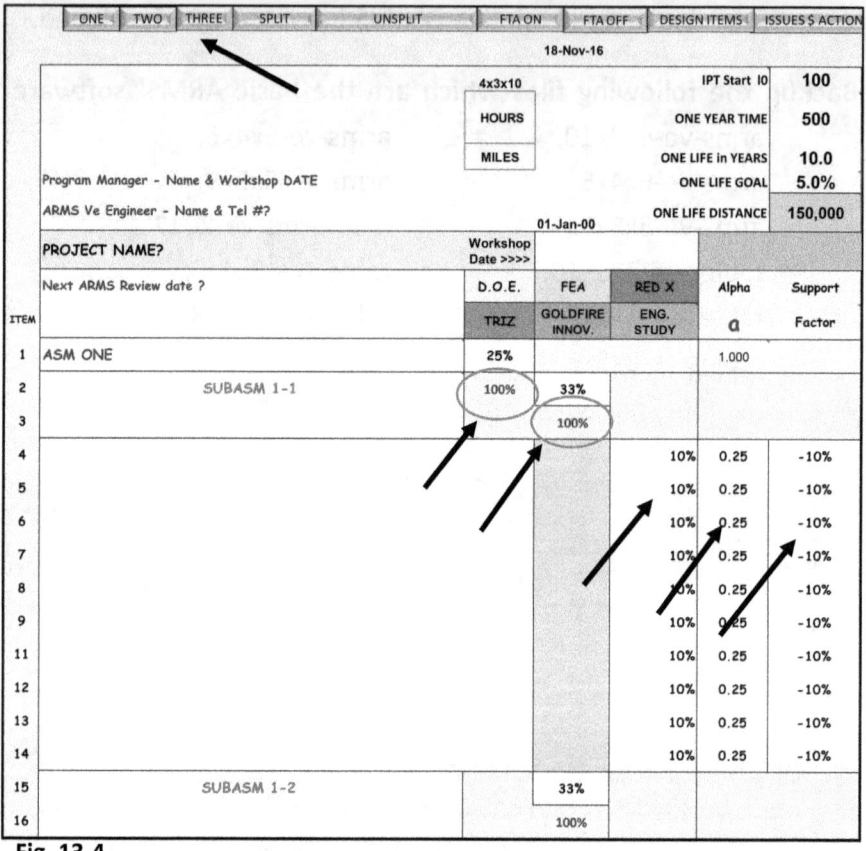

Fig. 13-4

ARMS. Press Update on the INPUT screen to form required links in Excel. If asked to enable macro's, press Enable.

SAVE your new model under a project name in the form:
arms-ve-project-name-yyddmm.

<u>ARMS' INPUT screen: Entries are made into white boxes only.</u>

Break the design down to fit one of the matrices. In this case, 4x3x10. Setting up a matrix is similar to making an indented list of drawings. Use button 'One' to enter 4 design segments. Allo-

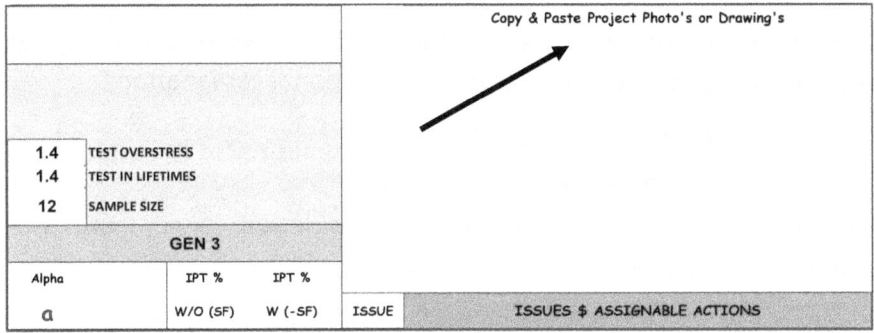

Fig. 13-5

cate risks which add to 100%. If you can enter any data above do so or wait to later when the required entries will be clearer. Capture discussions! Press button TWO and Enter, 3 each, design description's in the original four segments. Assign risks which add to 100% in each of the three. At any level, if a segment is not required, leave it blank. Press button THREE and add up to ten design descriptors in each of the 4x3 = 12 levels. This is up to 120

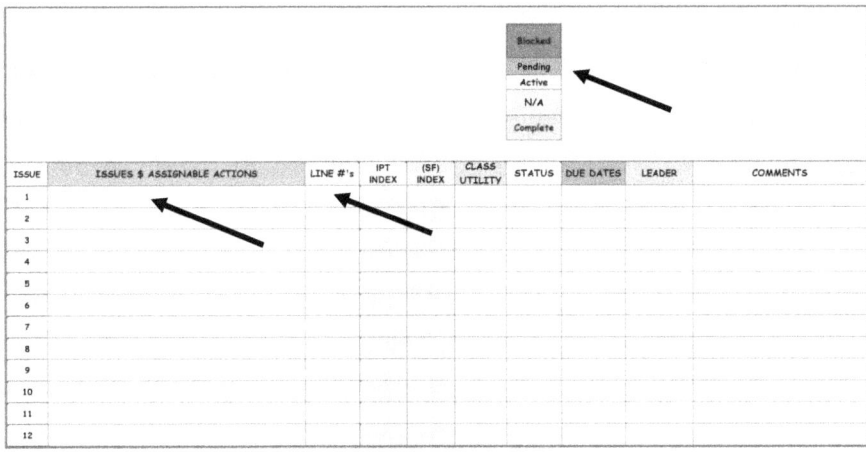

Fig. 13-6

design elements. Allocate risk to each of the ten that must add up to 100%. Capturing the discussions is paramount to the success of the ARMS' effort. Assign Alpha and Support Factors as needed. See guide sheet. By this time the team will be more familiar with the design and the boxes at the top of the INPUT sheet which should be filled in. Items requiring investigation e.g. Stress factor effects, should be assigned to an engineer to be resolved. Add Excel comments to any entry requiring additional explanation.

The use of comments is recommended, thanks to Dr. Ebbinghaus.

Insert a photo, sketch or diagram of the design concept into the right side of the INPUT sheet.

As Issues & Assignable Actions are discussed during the meeting, add them to the column for Issues & Assignable Actions with the line numbers from Column 'A'. If multiple line numbers are effected by the Assignable Action, make sure they are all listed in the box with a comma space format. **Capturing all affected line numbers is a critical step in the ARMS' report process.** These assignments do not affect the forecast but will affect the Pareto listing of Assignable Actions. Rows in the Assignable Actions column do not need to line up with the line numbers in column 'A'. The order of the items listed does not matter, because ARMS will Pareto sort the actions when the Action Item Report is generated. The other columns need to be filled in as the headers indicate. The Leader for an assignable action is being assigned the actions required to solve the design issue. Note that an Assignable Action may have a Support Factor in the same line This SF becomes an item for management review with the engineering leader and the project manager taking on the role of obtaining management's solution. Determining inputs for the boxes shown are a little more difficult to determine until after the team has been involved in several ARMS' Workshops. This is difficult at first, because some of the concepts are new to the

IPT Start	500	25	PROCESS % SIGMA				
ONE YEAR TIME	500	15,000	ONE YEAR MILES				
ONE LIFE In YEARS	10.0	1.2	OVERSTRESS FACTOR	1.3	TEST OVERSTRESS	1.4	TEST OVERSTRESS
ONE LIFE GOAL	5.0%	1.2	TEST IN LIFETIMES	1.3	TEST IN LIFETIMES	1.4	TEST IN LIFETIMES
ONE LIFE DISTANCE	150,000	6	SAMPLE SIZE	9	SAMPLE SIZE	12	SAMPLE SIZE

Fig. 13-7

team, but later it becomes fairly easy. IPT Start is a rough estimate of how many units would fail if one thousand were put into service for one year. In the beginning of the ARMS' Workshop, this value is merely a place holder. If the answer is "They would all fail," use a start value of 500 IPT. This interim place holder value will result in an ARMS' Weibull graph with slope values less than one as shown **(Fig.13-8)**. I think you would agree that with 500/1000 = 50% of the units failing in the first year, the design would be in a premature failure mode area. Note that the initial (top) line passes through 50% at the one-year time. Later we will have a better idea of this value and make adjustments to "calibrate" the ARMS' Weibull graph. The concept of %σ was explained earlier and needs to be entered. The default value is 33%. Discussions surrounding the setting of this value can be very revealing and issues need to be captured. A goal for the maximum number of failures after one lifetime of service needs to be set. Zero is not a goal. At this point do not set an unrealistic goal, be consistent with existing industry expectations. Negotiate with a customer if necessary, but the customer pays the bills. For electro-mechanical systems, we use 5% as a default, and for electronic systems 1%. In other industries, this goal may be considerably lower. Life support industries such as implants and safety of flight are examples.

Sample sizes are always a point of discussion. Early in the design effort, conformation test samples are expensive and you are not normally looking for the fine points of the design. In the early stage, it is a better use of budget funds to move fast with small samples and quick turnaround of design. The Trumpet Chart **(Fig.1-6)**, presented earlier, indicates that money put into obtaining nearly perfect nominal parts supports smaller sample sizes. If the source of parts has +/- 3 σ parts, that is not good enough. This is development work and we should seek parts that are approaching 10%σ. 10%σ parts allow samples of 3-4 and +/- 3 s parts should have samples of 9. Over 9 provides little useful in-

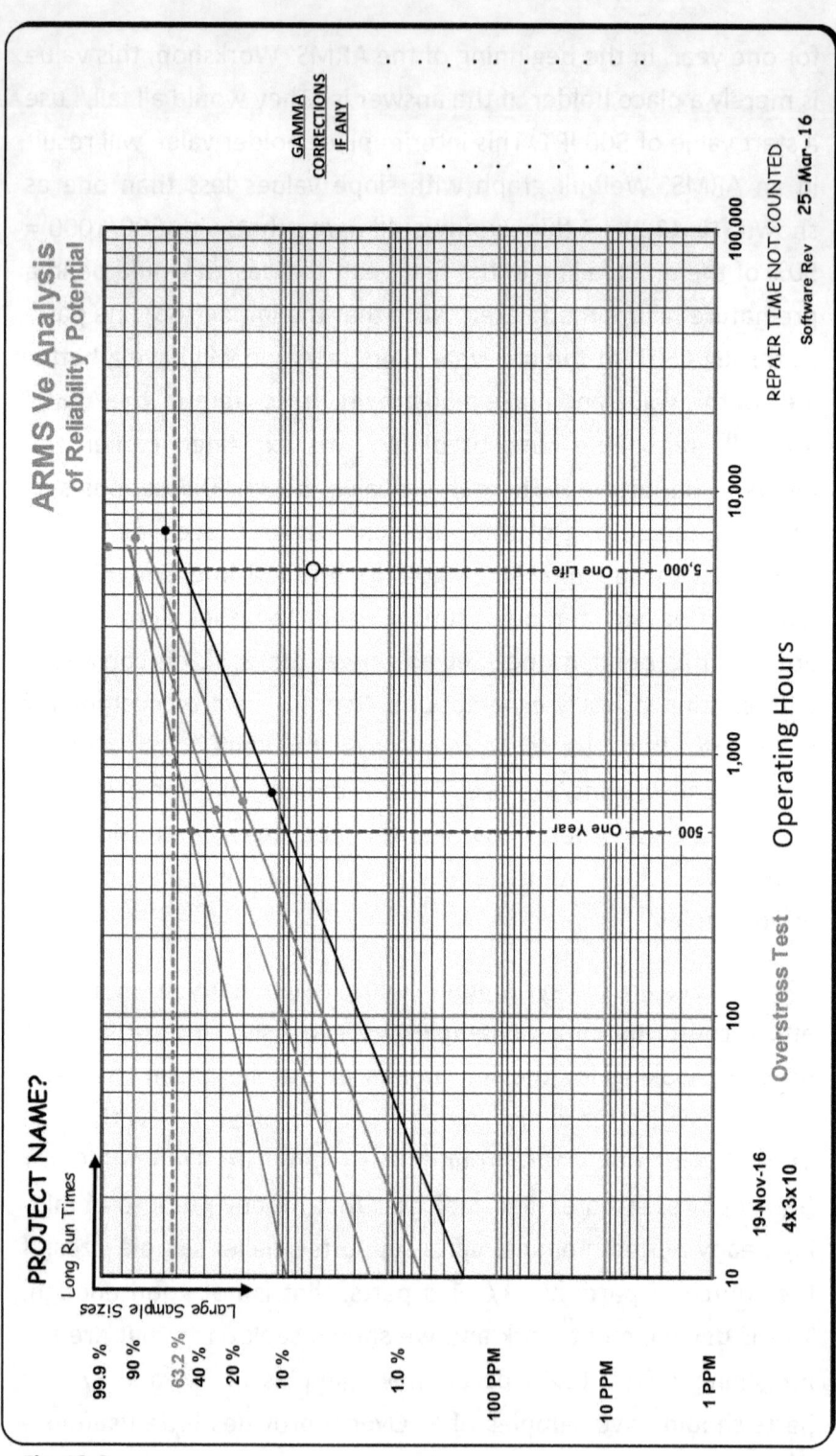

Fig. 13-8

280

formation at this point unless you are trying to prove your ability to make prototypes. That proof should come from capability studies of the processes

Chapter Fourteen

Making Weibull Graphs

- **To construct a Weibull graph, we will use tables of data that define and plot the horizontal and vertical lines that make up the Weibull grid.**

- **Facilitating Outstanding Leadership / Teamwork between all departments**

- **Engineering Design & Innovation Under Control**

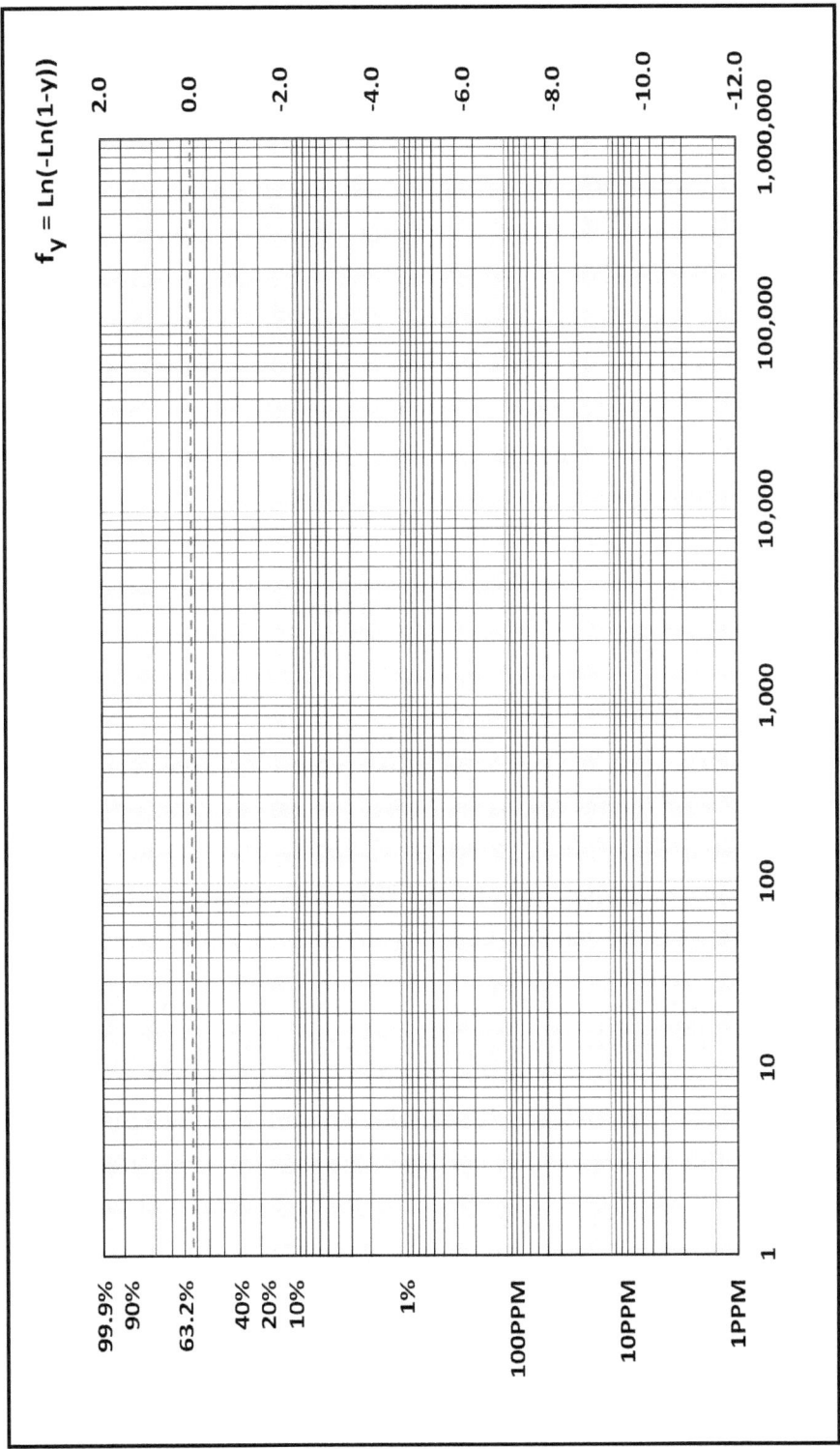

Fig. 14-1

283

Weibull Analysis using Ln(-Ln(1-y)) vs Ln(x) graphs have a unique abil- ity to display and clarify data from a wide variety of ap- plications. End of life data due to fa- tigue failures is one of the major appli- cations. To learn

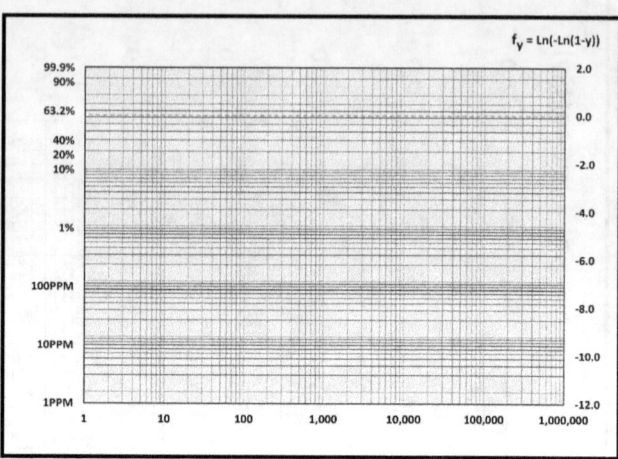

Fig. 14-2

more about Weibull applications, I suggest using an internet brows- er, searching the topic and looking for application books. Here, we will reveal a Weibull Graph construction approach and how to enter data into the Weibull graph.

The property of a Weibull graph that makes it of particular use is that data from a normal curve will plot as a straight line with a slope of one. The peak of the distribution will occur at a Ln(-Ln(1-y)) value of 0 which corresponds to 63.2%. This value is called the "Characteristic Life" and is shown as a dashed red line.

Weibull graphs are constructed in "cycles". The adjacent Weibull graph has 6 cycles on the vertical axis and 6 on the horizontal axis.

To construct a Weibull graph, we will use tables of data that define and plot the horizontal and vertical lines that make up the Weibull grid.

The full view of Table 1 is shown and to the left and right expanded to improve readability. The first column of Table 1, provides num- bers which will be used to identify 47 lines on the Weibull graph. A range of cycles .000001-.00001, .00001-.0001, .0001-.001, .001-.01, .01-.999 are converted to percentages using the Excel % format

1	0.00001	1	-11.5129	0.001%
2	0.00001	1,000,000,000,000	-11.5129	0.001%
3				
4	0.00002	1	-10.8198	0.002%
5	0.00002	1,000,000,000,000	-10.8198	0.002%
6				
7	0.00003	1	-10.4143	0.003%
8	0.00003	1,000,000,000,000	-10.4143	0.003%
9				
10	0.00004	1	-10.1266	0.004%
11	0.00004	1,000,000,000,000	-10.1266	0.004%
12				
13	0.00005	1	-9.9035	0.005%
14	0.00005	1,000,000,000,000	-9.9035	0.005%
15				
16	0.00006	1	-9.7211	0.006%
17	0.00006	1,000,000,000,000	-9.7211	0.006%
18				
19	0.00007	1	-9.5670	0.007%
20	0.00007	1,000,000,000,000	-9.5670	0.007%
21				
22	0.00008	1	-9.4334	0.008%
23	0.00008	1,000,000,000,000	-9.4334	0.008%
24				
25	0.00009	1	-9.3157	0.009%
26	0.00009	1,000,000,000,000	-9.3157	0.009%
27				
28	0.00010	1	-9.2103	0.010%
29	0.00010	1,000,000,000,000	-9.2103	0.010%
30				
31	0.00020	1	-8.5171	0.020%
32	0.00020	1,000,000,000,000	-8.5171	0.020%
33				
34	0.00030	1	-8.1116	0.030%
35	0.00030	1,000,000,000,000	-8.1116	0.030%
36				
37	0.00040	1	-7.8238	0.040%
38	0.00040	1,000,000,000,000	-7.8238	0.040%
39				
40	0.00050	1	-7.6007	0.050%
41	0.00050	1,000,000,000,000	-7.6007	0.050%
42				
43	0.00060	1	-7.4183	0.060%
44	0.00060	1,000,000,000,000	-7.4183	0.060%
45				
46	0.00070	1	-7.2641	0.070%
47	0.00070	1,000,000,000,000	-7.2641	0.070%
48				
49	0.00080	1	-7.1305	0.080%
50	0.00080	1,000,000,000,000	-7.1305	0.080%
51				
52	0.00090	1	-7.0127	0.090%
53	0.00090	1,000,000,000,000	-7.0127	0.090%
54				
55	0.00100	1	-6.9073	0.100%
56	0.00100	1,000,000,000,000	-6.9073	0.100%
57				
58	0.00200	1	-6.2136	0.200%
59	0.00200	1,000,000,000,000	-6.2136	0.200%
60				
61	0.00300	1	-5.8076	0.300%
62	0.00300	1,000,000,000,000	-5.8076	0.300%
63				
64	0.00400	1	-5.5195	0.400%
65	0.00400	1,000,000,000,000	-5.5195	0.400%
66				
67	0.00500	1	-5.2958	0.500%
68	0.00500	1,000,000,000,000	-5.2958	0.500%
69				
70	0.00600	1	-5.1130	0.600%
71	0.00600	1,000,000,000,000	-5.1130	0.600%
72				
73	0.00700	1	-4.9583	0.700%
74	0.00700	1,000,000,000,000	-4.9583	0.700%
75				

Fig. 14-3

function and placed into a column.

Moving down the second column of Table 1, the left and right ends of each ARMS' Weibull graph horizontal line will be defined utilizing two entries, and a blank space necessary to hide an Excel retrace line on the graph. The values in column two are set to cover our potential future needs for X values up to 10,000,000,000. High values of X are necessary when analyzing life time fuel injector cycles. The third column of Table 1 has 12 "Major Y Values", plus a line for the Weibull Characteristic Life. (63.2%). The fourth column is data for minor Weibull lines.

fy values of the major vertical axis.

$fy = \ln(-\ln(1-0.00001)) = -0.5192$

$fy = \ln(-\ln(1-0.00002)) = -0.8198$

$fy = \ln(-\ln(1-0.00800)) = -4.8243$

$fy = \ln(-\ln(1-0.99000))$
$= +1.5272$

#	X	Count	fx (a)	fx (b)	%
76	0.00800	1		-4.8243	0.800%
77	0.00800	1,000,000,000,000		-4.8243	0.800%
78					
79	0.00900	1		-4.7060	0.900%
80	0.00900	1,000,000,000,000		-4.7060	0.900%
81					
82	0.01000	1	-4.6001		1.000%
83	0.01000	1,000,000,000,000	-4.6001		1.000%
84					
85	0.02000	1		-3.9019	2.000%
86	0.02000	1,000,000,000,000		-3.9019	2.000%
87					
88	0.03000	1		-3.4914	3.000%
89	0.03000	1,000,000,000,000		-3.4914	3.000%
90					
91	0.04000	1		-3.1985	4.000%
92	0.04000	1,000,000,000,000		-3.1985	4.000%
93					
94	0.05000	1	-2.9702		5.000%
95	0.05000	1,000,000,000,000	-2.9702		5.000%
96					
97	0.06000	1		-2.7826	6.000%
98	0.06000	1,000,000,000,000		-2.7826	6.000%
99					
100	0.07000	1		-2.6232	7.000%
101	0.07000	1,000,000,000,000		-2.6232	7.000%
102					
103	0.08000	1		-2.4843	8.000%
104	0.08000	1,000,000,000,000		-2.4843	8.000%
105					
106	0.09000	1		-2.3612	9.000%
107	0.09000	1,000,000,000,000		-2.3612	9.000%
108					
109	0.10000	1	-2.2504		10.000%
110	0.10000	1,000,000,000,000	-2.2504		10.000%
111					
112	0.20000	1		-1.4999	20.000%
113	0.20000	1,000,000,000,000		-1.4999	20.000%
114					
115	0.30000	1		-1.0309	30.000%
116	0.30000	1,000,000,000,000		-1.0309	30.000%
117					
118	0.40000	1		-0.6717	40.000%
119	0.40000	1,000,000,000,000		-0.6717	40.000%
120					
121	0.50000	1	-0.3665		50.000%
122	0.50000	1,000,000,000,000	-0.3665		50.000%
123					
124	0.60000	1		-0.0874	60.000%
125	0.60000	1,000,000,000,000		-0.0874	60.000%
126					
127	0.63211	1		0.0000	63.212%
128	0.63211	1,000,000,000,000		0.0000	63.212%
129					
130	0.70000	1	0.1856		70.000%
131	0.70000	1,000,000,000,000	0.1856		70.000%
132					
133	0.80000	1	0.4759		80.000%
134	0.80000	1,000,000,000,000	0.4759		80.000%
135					
136	0.90000	1	0.8340		90.000%
137	0.90000	1,000,000,000,000	0.8340		90.000%
138					
139	0.95000	1	1.0972		95.000%
140	0.95000	1,000,000,000,000	1.0972		95.000%
141					
142	0.99000	1	1.5272		99.000%
143	0.99000	1,000,000,000,000	1.5272		99.000%
144					
145	0.99900	1	1.9326		99.900%
146	0.99900	1,000,000,000,000	1.9326		99.900%
147					

Fig. 14-4

$fy = \ln(-\ln(1-0.99900))$

$= +1.9326$

On the Weibull graph, Y vertical values will be positioned at horizontal logarithmic values utilizing the Excel Axis Formatting by checking the logarithmic box.

The same Weibull Graph can be used for a few cycles or several without re-programing (More on this later). To display X Ln values in the Weibull graph, check the Excel logarithmic function box on the X axis format screen. Major axis lines may be individually formatted red and the line width adjusted.

$X = 1$, $f_x = \ln(X) = 0.00000$

$X = 2$, $f_x = \ln(X) = 0.6931$

$X = 10$, $f_x = \ln(X) = 2.30259$

$X = 100$, $f_x = \ln(X) = 4.60517$

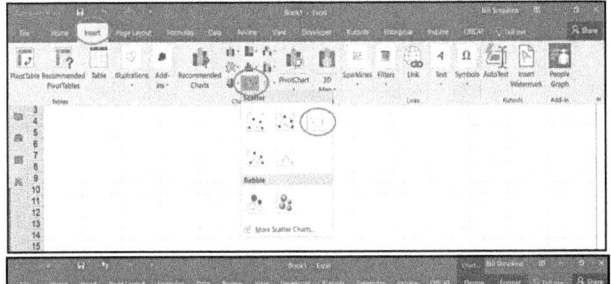

(1) Open a new worksheet with Windows and insert a new graph using the X-Y scatter type without smooth lines.

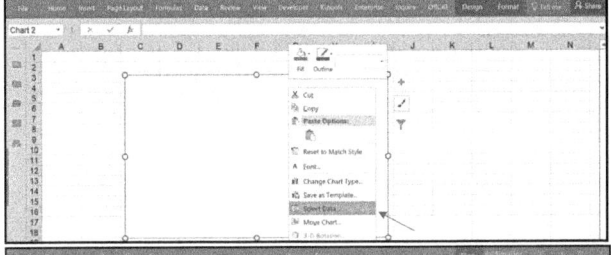

(2) Place the cursor on the open graph area and right mouse click then open a window of "Chart Tools." Click on "Design" and "Select Data."

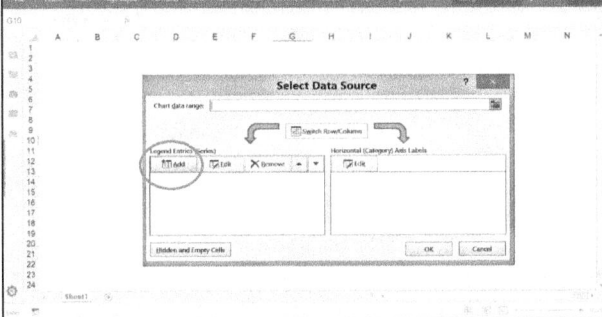

(2) Select "Add," a window "Edit Series" will open. Fill in the "Series name" for X axis label as "Weibull Analysis" then press OK. Click on the "Series X values" box then go to the Work Sheet containing your

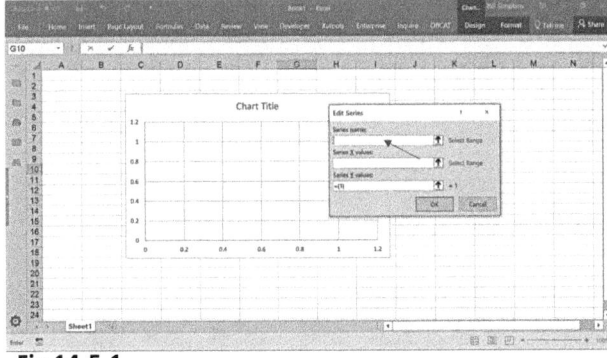

Fig.14-5-1

columns of Weibull lines data (Fig.14-3,4) then highlight the entire third column for Major X axis values then press "Enter."

(4) Click on the "Series Y values" box then go to the Work Sheet containing your columns of Weibull lines data and highlight

287

the entire second column for the Y axis then press "Enter."

(5) Repeat steps (3) to (5) for the fifth column "Minor X axis lines" and the fourth column "Characteristic Life."

(6) For the vertical lines, highlight the X axis on the graph, "Select format X axis," check the Logarithmic scale box. Check Major axis lines and Minor axis lines boxes. Format line colors as desired. Red major and black minor is a common standard.

(7) Right mouse click in the center of the plot and format the plot area. A light yellow, from custom colors, for the plot area works well.

(8) Highlight the horizontal axis and "Format axis" to adjust decimal points to zero and set font size to your liking. Repeat for vertical axis.

(9) Set Axis crossing values for both axis. 1.0 and −12.0 will set the vertical axis to the right.

(10) Manually add percent labels to the left side of the graph for major lines. Values are found in the columns of data. The vertical axis values are labeled as "Percentages & Parts Per Million" and the horizontal in cycles of interest. e.g. Hours, miles, cycles of operation, etc. With logarithms you cannot plot 100%. The maximum value is just short of 100%. e.g. 99.9%.

(11) Size the plot area to allow a generous border for titles, dates, etc.

The left-hand scale is $Ln(-Ln(1-y))$ and the horizontal scale is $Ln(x)$. Normally, each tenth line is color coded and thicker for clarity. In this example, red with the intervening values as thinner black lines.

The Ln(-Ln(1-y)) values on the right side of the graph may be hidden using the axis format screen.

Save your work in a separate file labeled "Weibull Screen."

To enter data into the ARMS' Weibull graph:

1) make a copy of "ARMS' Weibull Screen" and "save as" with a new name of your choice.

2) Place curser on the graph, right click and select "select data."

3) Select "Add" and enter series title, series X and series Y information in the normal Excel method. Vertical axis values are ln(-ln(1-Y)) and Horizontal axis values are ln(X).

4) Save.

5) Add a new sheet and build a table to enter data and calculate median ranks. The following is from the ARMS' program;

Three data points are needed for each line,

(1) quantity taken off test, failed or suspended, and

(2) hours / cycles / miles on test. Units still on test are reported as though the test program was suspended, this provides a snap shot status. Actual testing continues uninterrupted. If the units

| F6 | | | | × ✓ ƒ× | =IF(D6<0.5,IF(C3-SUM(C6:C6)>0,((C3+1-MAX(G5:G6))/(1+H6)),0),1) | | | | | | | | |

List suspensions before failures if times are equal.

(Test 1)			γ Gamma Correction =		Program Name					Sample Size 95% Conf (+)	Sample Size 95% Conf (-)	
TOTAL	10					Beta Est.1 >>		1	670			
Line #	QTY TAKEN OFF TEST	Failure1 Suspension0	Rank Individual TIMES or CYCLES at Test Termination	INCREMENT	Mean Order Number	NIBPSS	Median Rank %	Plot Point Y	Plot Point X	"Applied Life Data Analysis" Wayne Nelson	"Applied Life Data Analysis" Wayne Nelson	Comments
			HOURS		1.11					+ θ	− θ	
1	1	1	400	1.000000	1.000		6.897%	-2.6691	400			
2	1		450	1.111111		8.00		-2.6691	400			
3	1	1	490	1.000000	2.111		17.389%	-1.6553	490			
4	1		500	1.269841		6.00		-1.6553	490			
5	1		530	1.481481		5.00		-1.6553	490			
6	1	1	589	1.000000	3.593		31.646%	-0.9664	589			
7	1	1	670	1.000000	5.074		46.901%	-0.4872	670			
8	1		700	1.976309		2.00		-0.4872	670			
9	1		730	2.962963		1.00		-0.4872	670			
10	1		750					-0.4872	670			
11												
12												

Fig. 14-8

289

	List suspensions before failures if times are equal.							
	TEST 1		$\Gamma =$			Program Name		
TOTAL	10						Beta Est.1 >>	
Line #	QTY TAKEN OFF TEST	Failure=1 Suspension=0	Rank Individual TIMES or CYCLES at Test Termination	INCREMENT	Mean Order Number		NIBPSS	Median Rank %
			HOURS		1.11			
1	1	1	400	1.000000	1.000			6.697%
2	1		450	1.111111			8.00	
3	1	1	490	1.000000	2.111			17.389%
4	1		500	1.269841			6.00	
5	1		530	1.481481			5.00	
6	1	1	589	1.000000	3.593			31.645%
7	1	1	670	1.000000	5.074			45.901%
8	1		700	1.975309			2.00	
9	1		730	2.962963			1.00	
10	1		750					
11								

Fig. 14-9

that are running are taken off test for performance evaluation, you will need a gamma value equal to the run time completed to correct the Weibull presentation.

Total quantity on test is automatically calculated.

Total= SUM (C6:C30)
INCREMENT=
IF(D6<0.5,IF(C$3-SUM(C$6:C6)>0,((C$3+1-G5)/(1+H6)),0),1)

Place in first INCREMENT cell, then drag cell down the column. In a similar manner, the individual cells in ARMS may be highlighted and the formula for that cell reviewed and copied for pasting into your new sheet. Because of suspended sets that may occur in your data, the strategy for assigning formulas has been set to correct for the gaps in data. Of course, you may use a copy of ARMS to separately evaluate any data In the ARMS' model. There are several example data sets to the far right of the data entry sheets that may be applied to verify that your calculations provide the correct answers.

Appendix

3D Warranty Data.

A Vision of a Coffee Cup.

- With the 3D Warranty Graph, it is clear if reliability has improved or deteriorated in the near term.

- With ARMS' 3D analysis graphs the reliability engineer's mind is an active part of the data reduction process.

- ARMS is about engineers forming a vision and using the laws of nature to bring it into a reality.

- Facilitating Outstanding Leadership / Teamwork between all departments

- Engineering Design & Innovation Under Control

- Enjoy your cup of coffee as you ponder what does it all mean.

Several examples of decomposing data using Weibull graphs in a GUI approach were presented earlier. The 3D graph approach is used to analyze massive data sets looking for patterns in the data that should receive closer attention.

During fifteen years in the HVAC Industry, the first Reliability question asked was always, "How many units will be returned in the next quarter or two?" Once estimated, the next question was, "What needs to be done to improve our results?" Answering the second question led to the development of a "3D Warranty" methodology that allows analysis of massive amounts of data sensitive to small anomalies.

The first step in a 3D Warranty analysis is to "Normalize" the data to the month of Manufacture. This will place sales in the first column and the initial returns from each sales month in the second

Manufacture / Sales Date	Sales Qty	NORMALIZED DATA FOR MONTHS IN SERVICE															
		1	2	3	4	5	6	7	8	9	10	11	12	13	14	15	16
Apr-15	6500	127															
Mar-15	7500	123	95														
Feb-15	9000	111	41	55													
Jan-15	10000	26	46	27	35												
Dec-14	10000	7	34	54	15	25											
Nov-14	9500	8	43	47	26	21	15										
Oct-14	11000	6	25	54	37	28	43	25									
Sep-14	9000	5	32	25	16	34	18	32	25								
Aug-14	10000	7	31	44	32	24	16	14	36	31							
Jul-14	11000	3	28	46	30	22	34	18	64	81	76						
Jun-14	10000	101	32	45	64	37	34	82	187	221	266	295					
May-14	9000	99	37	55	72	43	33	45	111	194	227	289	315				
Apr-14	8500	89	36	53	86	92	21	33	74	163	187	228	317	335			
Mar-14	7500	88	39	54	54	31	82	36	42	68	143	162	247	256	305		
Feb-14	7000	86	42	45	42	34	33	76	31	55	116	145	263	277	314	295	
Jan-14	6500	78	43	40	35	25	22	15	85	45	85	125	225	265	305	285	295

Fig. App.-1

column as shown. The columns are months since sale. Each horizontal line of data represents the history of returns from a particular month of manufacture by serial number. If manufacturing data is not available in sales data, use sales numbers and we will deal with the loss of clarity later.

The "Normalized" return data is now matched to the sales (sales / manufacture) month from which it came.

Month																
Apr-15	2.0%															
Mar-15	1.64%	1.27%														
Feb-15	1.23%	0.46%	0.61%													
Jan-15	0.26%	0.46%	0.27%	0.35%												
Dec-14	0.07%	0.34%	0.54%	0.15%	0.25%											
Nov-14	0.08%	0.45%	0.49%	0.27%	0.22%	0.16%										
Oct-14	0.05%	0.23%	0.49%	0.34%	0.25%	0.39%	0.23%									
Sep-14	0.06%	0.36%	0.28%	0.18%	0.38%	0.20%	0.36%	0.28%								
Aug-14	0.07%	0.31%	0.44%	0.32%	0.24%	0.16%	0.14%	0.36%	0.31%							
Jul-14	0.03%	0.25%	0.42%	0.27%	0.20%	0.31%	0.16%	0.58%	0.74%	0.69%						
Jun-14	1.01%	0.32%	0.45%	0.64%	0.37%	0.34%	0.82%	1.87%	2.21%	2.66%	2.95%					
May-14	1.10%	0.41%	0.61%	0.80%	0.48%	0.37%	0.50%	1.23%	2.16%	2.52%	3.21%	3.50%				
Apr-14	1.05%	0.42%	0.62%	1.01%	1.08%	0.25%	0.39%	0.87%	1.92%	2.20%	2.68%	3.73%	3.94%			
Mar-14	1.17%	0.52%	0.72%	0.72%	0.41%	1.09%	0.48%	0.56%	0.91%	1.91%	2.16%	3.29%	3.41%	4.07%		
Feb-14	1.23%	0.56%	0.60%	0.56%	0.45%	0.44%	1.01%	0.41%	0.73%	1.55%	1.93%	3.51%	3.69%	4.19%	3.93%	
Jan-14	1.20%	0.66%	0.62%	0.54%	0.38%	0.34%	0.23%	1.31%	0.69%	1.31%	1.92%	3.46%	4.08%	4.69%	4.38%	4.54%

Fig. App.-2

The second step, calculate the non-cumulative percent return for each month in service for each month of sales.

The third step, develop a three-dimensional plot.

(This example was developed with MS EXCEL.)

Each slice of data is the "in warranty return" from a specific sales month quantity. (S/N returns vs Production data would be better, but the data was not available.) See the following graph. The blue slice of data for Jan-2014 had a 1.0% peak failure in the month after sale. Another peak in failure occurs later in life around the 11th month after sale. Note that the data points are not averages or accumulations; there is only one point in time when Jan-2014 product is one-month old. The Mar-2014 product experienced approximately 4% failure after 13 months in service.

The diagonal Red dashed line represents today, the mountainous area to the left is failure return history and the blank area to the right is in the future. The return history to the left does not grow! It is "there" or "not there." The 3D Warranty Graph is displaying the "real time instantaneous failure rate!" Each month, mountains at their final height will pop-up to the right of the diagonal, while the mountains to the left will remain unchanged. I actually made a solid model from real data, using Lexan sheets of plastic that were edge light. The model was used for customer presentations to show where and how improvements were being made. The data in this example is illustrative and not from actual results.

Learning and using specific industry characteristics are part of the

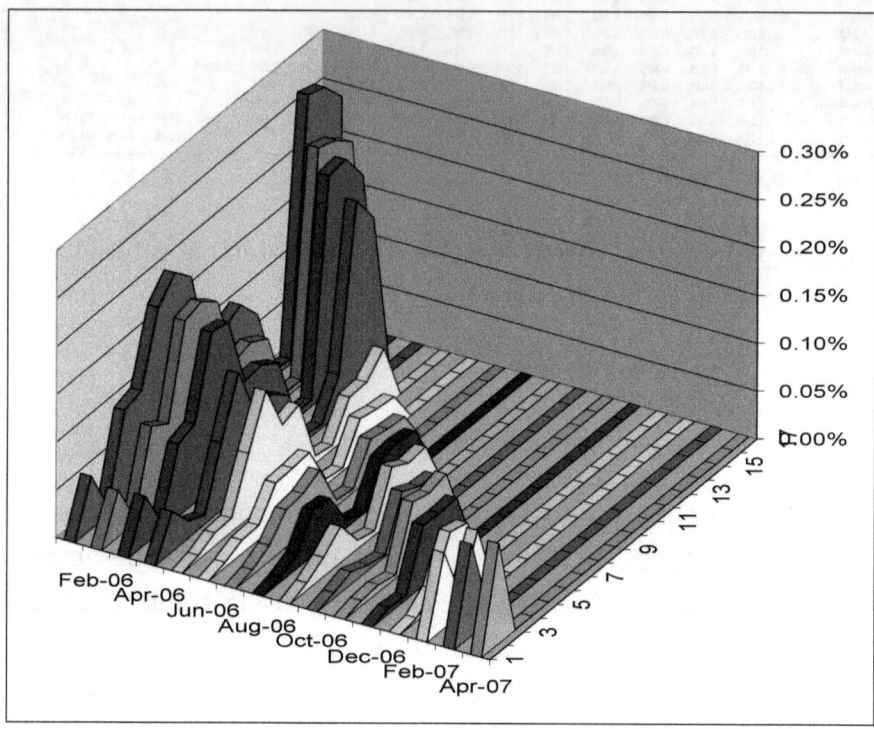

Fig. App.-3

required analysis. Knowing what time lines in the field occur along a diagonal line allows assessing the possible seasonally induced effects. In this example, there is a seasonal pattern. Seasonal patterns common to the HVAC industry exhibit strong waves that form peaks and valleys along a diagonal. In the HVAC industry, distributors are busy installing product leading up to the spring-summer cooling season. During the fall, they are busy returning failed product. This makes it important to capture the date of actual failure instead of the date of return.

Customers track failure results by matching seasonal returns to yearly purchases. They typically do not match serial numbers. In the 3D Warranty Graph, they are looking along the diagonals and getting an average percent of failure from sales and returns of different manufacturing dates. The compressor supplier is looking parallel to the slices at manufacturing date and serial number matched returns from those dates of manufacture. A customer will be comparing older second year failures to current year purchases. The customer is blind to improvements made in the current year.

With the 3D Warranty Graph, it is clear if reliability has improved or deteriorated in the near term.

In the data presented, you will note that starting in Jun, Jul and Aug, the mountains are getting lower and by Sept they are gone. This is due to real improvement in the design, manufacture and application. If the product was the same as earlier months of manufacture, the mountains would be along the left edge of the full length of the diagonal between the past and future.

Because customers do not have the ability to normalize the data, they look at a band of data parallel to the diagonals, usually one year wide. The real dramatic improvement that occurred will remain obscured to the customer until older failures are out of his range. You can prove that improvements have been made even though the customer does not see them yet due to the way customer data is collected and displayed.

Reading the 3D Warranty Graph

Failures during the first three months in service occur across the front rows 1-3 and are classified as from infant mortality causes. Dead on arrival and miss-applications are examples of how this can occur. Each particular industry will have characteristics that are general to this region. Failure analysis information on infant mortality products will provide insight to specific causes.

If products are often returned with nothing found wrong, the problem is usually influenced by the distribution system return policy as much as the product itself, but caution, there are many true causes that can lead to infant mortality.

Changes in infant mortality can demonstrate improvements in manufacture or implementation of design change improvements. Improvements / degradations are quickly noted. Classical moving average data obscures changes in reliability. The 3D Warranty Graph is instantaneous and not an average.

(Jan-Jul)-2006 returns during the first two months service indicate a high initial failure rate and may be an early indicator of problems in the product. Rapid response to analyze early returned products is needed. In this case, a cause was found and corrective actions applied. Returns from (Jul-2006 to Jan 2007) are greatly improved. Note the early returns in (Feb-Mar-Apr) 2007. With the typical averaging of warranty data, it would take several months to recognize a problem in the installed units.

Data entered into the 3D Warranty Graph can be filtered to only include the results from a single customer and be used for customer support. This was done often. In one case like the above data, an all customers model revealed an early failure problem for a specific model HVAC compressor sold to several customers.

After data mining by filtering, the problem was quickly isolated to one system manufacturer. One distributor was also having an unusually high return of another compressor model. The Failure Analysis Technicians were alerted to give priority to these early returns.

Failure analysis of the specific customer's returns identified the failure cause, low refrigerant oil. During a review with the customer, who had no idea a problem was developing, it was discovered that the customer made a system change effecting the normal return of refrigeration oil that circulates in the closed loop system. Compressors were being starved of oil and eventually failing, some quickly. With this information, the customer was able to make an immediate change in production to correct the issue. Quick response to apply field service corrections saved the customer several hundred thousand dollars.

The single distributor was near Dallas, Texas. The distributor was installing heat pump units in tract homes as they were being built. To investigate, a trip was made to the job site. It was a massive building effort, over one hundred homes were being built at a time with thousands scheduled for construction. Work was conducted by roving teams. One team poured foundations, another framed, another applied plywood siding, etc. HVAC had teams to pour concrete pads, another set units, a third connected the Freon lines to the home at a common service point which contained all services, electric, water, and HVAC. A fourth team member connected the wiring and ran the unit for a few seconds to confirm it was ready for the next team. The fifth team member applied a Freon charge and ran the unit long enough to confirm that it could provide cooling and heating. Talking with the supervisor revealed that some team members were unskilled, untrained day employees who were instructed in what to do each morning. It turned out that at times, the unit Freon lines were connected to the house water lines and not the evaporator Freon lines. The plumbing team member, at times, would inadvertently turn on water and fill the HVAC unit. When that happened, the plumber would fix the error and say nothing, not knowing a problem was created. The HVAC technician would come later, charge the unit with Freon and run it, but not long enough to see the water problem. It would go undetected until a home owner was on the scene. Then a compressor would be returned from a reported month or two of service. All of this came to light by analyzing returns with a 3D Warranty Graph. Again, thousands of dollars were saved. Conventional warranty analysis would never have been able to recognize the problem. Correction only came from a Failure Analyst being able to recognize a problem and make inquiries and go to the job site when necessary. In another case, a spike in returns out of season identified a distributor who was falsely reporting failures in a program that allowed returning metal serial number labels only, saving money on compressor freight. During a review of their facility, a machine was discovered for making counterfeit labels.

In another case, a spike occurred due to 50 compressors returned out of season. The cause was compressors from an engineering field trial that were removed from service at the end of the trial. Later the local distributor returned them for warranty credit! The distributor returned the warranty funds when it was brought to his attention.

The pattern of failures indicates infant mortality followed by useful life, and as the first-year service is approached, larger failure rates are reported. It is not a coincidence that the warranty period is one year. Failure analysis of returns in months 11-13 after sale aid in determining where corrective actions are needed in design and / or distribution.

A comparison between customer "A" and all other customers for one specific model proved pivotal in convincing the customer that they had an application problem.

This ability to provide early accurate results can assist in obtaining additional business and in alerting customers early of problems on the horizon. This is strong support when building or maintaining a relationship.

Additional steps to create a series of 3D Warranty Graphs by filtering the input data make the 3D Warranty Graph even more useful. For example, filtering to specific customers, distributors, application types, voltages, or geographical regions or other characteristics common to your industry. The 3D graph shown was built with Excel, the original work was built using SAS which supports handling large data sets. A 3D graph approach requires large data sets and will not work well with limited data.

With most data mining software, you crunch a bunch of numbers following an algorithm and obtain a numerical result.

With ARMS' 3D analysis graphs the reliability engineer's mind is an active part of the data reduction process.

This involvement occurs as the 3D wave form is visually reviewed for anomalies worthy of receiving focused detailed analysis.

A Vision of a Coffee Cup

ARMS is about engineers forming a vision and using the laws of nature to bring it into a reality. Before departing, I would like to discuss another vision. Your cup of coffee or, if you prefer, your cup of tea. It is the cup I am focused on, you can drink the rest. How is the cup a vision? Somebody thought it up and formed some dirt (clay) to give it a form as visualized. They fired it in an oven to transform the clay into a ceramic. The ceramic is a crystalline structure formed by the clay making cross linking bonds. Perhaps it was flow coated with powered glass and re-fired to make a surface that is pleasing to the eye. The result is a "real" cup, we can feel it, bang it on the table and hopefully drink from it.

But what is really going on with the cup? Let us look closer, let us use our ability to form visions. As our minds eye approaches the cup, we come closer and closer until we begin to see that the cup has a structure made up from the glass molecules of the material. We can see beyond the surface structure of the glass and see crystals of the ceramic. There are wide spaces between the ceramic molecular structure. Yet thanks to the force fields around the glass molecules, our coffee cannot slip into the ceramic structure. If we had broken the cup and exposed the raw edges of the ceramic to the coffee, we would see that the coffee, in fact, does penetrate the ceramic structure, leaving a stain. We are still able to take our mind's eye in deeper until we discern the presence of atomic level activities. At this level, we recognize that the relationship between atomic items and the spaces between them is vast. We are into a realm of a void with electro-magnetic fields and very little solid material, if any. Our cup is mostly a void by a wide margin. Keep going, deeper and deeper. The atoms are made of sub-atomic entities, quarks, masons, bosons, etc. I believe at least seven have been identified. Closer and closer, deeper and deeper, we come to the "Higgs boson" believed to be the source of gravity! Closer yet, present theories widely accepted would have us find photons and force fields. This is as far as we will go with our trip. At this point, again the major component is a void. Our cup is a void with some photons at the lowest level. How real is that? Is our cup an illusion? I can bang it on the table or my head for that matter. Of course, finding a void there would not surprise anyone:)

The cup is real because that is how we define reality. We look at the things around us that we can see, feel and touch. But our science is expanding our abilities and making greater use of our ability to visualize beyond our sight and feel. I worked on building mirrors for telescopes that have looked into deep space and formed visions of galaxies over 50 million light years away. That is a lot of void in between. Now they are working to build instruments that will double that distance. We expect to find more physical bodies at that massive 100 light year distance. Perhaps visions of the original start of all of this.

Where do you think all of this started? How did it start? What is the original source? What shall we call it? I think I know the answer. However, I will let you figure it out for yourself. Think of it as homework. **Enjoy your cup of coffee as you ponder what does it all mean.** I hope that you can agree with me. I usually have a second cup to enjoy as I think about the first one.

About the Author

Bill Simpkins and ARMS Ve

A Synergetic Process for Innovation Development

Bill Simpkins has over 40 years of experience working as an Associate Engineer at the Bell Telephone Laboratories , as a Project Engineer, Engineering Manager, Reliability Manager, Director of Quality, Consultant to NASA and the National Reconnaissance Office, and at GM / Delphi as an Advanced Global Reliability Manager. During this period, Bill earned degrees in Electrical Engineering, Physics and an MBA. These opportunities led to a close association with numerous engineering teams throughout North America, Europe, and parts of Asia. Examination of what engineers require to be successful and systemic threads of issues that inhibited them from achieving their objectives led to the development of ARMS.

ARMS is a management system structured to improve engineering team performance, resource utilization and product reliability, while documenting what the engineering team believes, at several stages, as the design evolves.

ARMS a GUI interface between Engineering and Statistics. If there are missing essential elements in your design process, ARMS will bring this fact to the table and permit developing solutions earlier in the process than would occur with conventional approaches. I have not yet met an engineering team that ARMS did not improve. Specialized engineering tools that are add-ins to the design process can be justified and scheduled before budgets are exhausted, e.g., Design of Experiments, Finite Element Analysis, bench tests, training programs, etc.

This handbook on how to apply ARMS is not intended to be a statistical textbook. The ARMS' Handbook is focused on how to apply

ARMS to an advanced engineering effort, reducing risks and reaping the benefits of GUI (Graphical User Interface) techniques.

Universally, it is normal for silos of power to develop within an engineering company. Management, Finance, Purchasing, Manufacturing, Engineering, Reliability and Quality often are in a posture to defend their own turf, particularly when budgets are involved. Company bonus structures exacerbate any points of friction. Bonuses reward getting the products out under budget and on time. However, this is a double edged sword, because product designs released to production before they are ready also reap rewards. During product development, reliability or risk reduction is not measured for employee reward. Most decisions affecting reliability and risk reduction are made out of sight by a limited number of people. And risk decisions are often made by unqualified individuals without conscience knowledge that they are making a risk decision. Decisions bad for risk usually come in the form of limiting resources and time. It is common to rush to production before durability tests have demonstrated that required reliability has been or may be achieved.

ARMS pulls the company administration and design team together to make key decisions visible and transparent.

ARMS speeds up design cycles and usually results in more design turns for less money and less time, to the benefit of product development.

ARMS provides a structure for the newer team members and less experienced engineers to learn and become a greater asset on future projects. Any engineering team has far more talent and experience than is evident or currently being utilized.

ARMS discovers hidden talents and pulls them into the design equation through open interactions.

ARMS provides education for the engineering and management team on what is important during advanced design development and what can be left to another time.

ARMS reduces waste and converts what would otherwise be waste, into resources.

ARMS identifies supplier strengths and weaknesses while they can be addressed, before supplier management failures occur.

ARMS supports, when permitted by your management, the customer in becoming a participant in your design process, thus forming a stronger bond with your company.

This ARMS' Handbook was compiled to demonstrate how ARMS can support risk reduction and innovation.

ARMS is a process designed to lead the development team to making correct decisions.

ARMS is not a statistical exercise. The required math has been pushed into the background, allowing non-statistically trained team members to make solid contributions and guide the team in taking the actions necessary to succeed.

This ARMS' Handbook is meant to be a reference for engineers engaged in ARMS' Workshops. It should be reviewed prior to an ARMS' Workshop, be available during the workshop and referenced as necessary. ARMS' software, available from Bill Simpkins, is required to conduct an ARMS' Workshop.

In the ARMS' software there is a file, "load-arms-ve-2017," with a main sheet containing a number of tabs at the bottom. Reviewing sheets connected to these tabs will demonstrate concepts important to ARMS.

Normally, an ARMS' educator is required to guide the ARMS' process until the team becomes knowledgeable and comfortable with ARMS. The ARMS' educator will review these concepts at the start of a new ARMS' Workshop and whenever required thereafter.

A chapter on "Making ARMS' Weibull Graphs" was included to answer questions on how the Weibull Graphs were constructed. Knowing how to build Weibull graphs is not necessary to the execution of an ARMS' Workshop and will only be reviewed on request.

Appendix "3D Graphs" is an add-on and not a part of ARMS. The 3D Graph can be a powerful tool if you are dealing with large quantities of field data. The 3D Graph was developed to evaluate millions of HVAC compressor field returns and be sensitive to problem trends early in their development.

References

Irving W. Burr
Specifying the Desired Distribution Rather than Max and Min Limits
Industrial Quality Control Aug 1967

Eugene L. Grant & Richard Leavenworth
Statistical Quality Control
ISBN 0-07-024114-7

John D. Kalbfleish & Ross L. Prentice
The Statistical Analysis of Failure Time Data
ISBN 0-471-05519-0

Dimitri Kececioglu
Reliability Engineering Handbook   Vol 1 & 2
ISBN 0-13-772294-X

Reliability & Life Testing Handbook Vol 1 & 2
ISBN 0-13-772369-5

Dimitri Kececioglu & Feng-Bin Sun
Environmental Stress Screening
ISBN 0-13-324229-3

K.C. Kapur & L.R. Lamberson
Reliability in Engineering Design
ISBN 0-471-51191-9

J. F. Lawless
Statistical Models and Methods for Lifetime Data
ISBN 0-471-08544-8

Elisa T. Lee
Statistical Methods for Business and Economics
ISBN 0-534-97987-4

Wayne Nelson
Applied Life Data Analysis
ISBN 0-471-09458-7

Weibull Analysis of Reliability Data with Few or No Failures
Journal of Quality Technology Vol 17 No.3 July 1985

R.C. Plaffenberger & J.H. Patterson
Statistical Methods for Business and Economics
ISBN 0-256-92350-6

S. Timoshenko & D.H. Young
Elements of Strength of Materials   D.Van Nostrand Company, Inc.

Anthony Rizzi

The Science before Science   IAP Press
ISBN 1-4184-6504-6

William J. Simpkins Sr.
Potentiometer Failure Analysis Technical Report # 351-74-466
National Reconnaissance Office   Project 9632

ARMS ILLUSTRATIONS (Fig.#)

www.ingramcontent.com/pod-product-compliance
Lightning Source LLC
Chambersburg PA
CBHW071411180526
45170CB00001B/67